SUSTAINING EARTH
Pathways to Resilience

About the Editors

Banshaikupar Lyngdoh Mawlong is as an Assistant Professor of Political Science at Union Christian College in Meghalaya, India. His primary area of specialization is Environmental Governance. He has authored several research articles in academic journals, focusing on topics such as environmental governance, cultural sustainability, environmental politics, and women's participation in politics. He has edited four volumes: *Environment-Cultural Interactions and the Tribes of North East India* (2015), *Climate Change and Developing Countries* (2018), *Culture and Sustainability* (2019), and *Women & Development* (2020).

Jonestar L. Nonglait is an Assistant Professor in the Department of Geography at Union Christian College, Umiam Khwan, Meghalaya. He has been actively engaged in teaching, research, and academic coordination. His research interests span environmental governance, cultural sustainability, and watershed management and he has published in reputed journals and contributed to edited volumes.

SUSTAINING EARTH
Pathways to Resilience

Edited by

Banshaikupar Lyngdoh Mawlong
Jonestar L. Nonglait

CONCEPT PUBLISHING COMPANY PVT. LTD.
NEW DELHI - 110 059

ISBN: 978-93-6344-436-2

First Published 2026

Published and Printed by

Concept
Since 1974

Concept Publishing Company Pvt. Ltd.
A/15-16, Commercial Block, Mohan Garden
New Delhi-110059 (India)
T: +91-11-41101460
E: publishing@conceptpub.com, *W*: www.conceptpub.com

Cataloging in Publication--Courtesy: D.K. Agencies (P) Ltd. <docinfo@dkagencies.com>

International Seminar, Earth on the Edge: Tackling Global Environmental Crisis (2025: Shillong, India)
Sustaining earth : pathways to resilience / edited by Banshaikupar Lyngdoh Mawlong, Jonestar L. Nonglait.
pages cm
"... International Seminar, Earth on the Edge: Tackling Global Environmental Crisis, organised by Union Christian College on February 12 and 13, 2025"--Preface.
Includes bibliographical references and index.
ISBN 9789363444362

1. Environmental degradation--India--Congresses. 2. Environmental degradation--India--Meghalaya--Congresses. 3. Sustainable development--India--Congresses. 4. Sustainable development--India--Meghalaya--Congresses. 5. Environmental protection--India--Congresses. 6. Environmental protection--India--Meghalaya--Congresses. I. Mawlong, Banshaikupar L., editor. II. Nonglait, Jonestar L., editor. III. Union Christian College (Shillong, India), organizer. IV. Title.

LCC TD171.5.I4I58 2025 | DDC 363.700954 23

Foreword

In a world where climate shifts are no longer predictions but present realities, the urgent need to confront our environmental challenges has never been clearer. The present anthropogenic environmental crisis has reached unprecedented levels of urgency and impact, requiring the immediate attention of all earthlings. In the past decade, however, these challenges have become more acute. Scientific data increasingly shows that the Earth's systems are being pushed beyond their natural limits, resulting in more frequent and severe extreme weather events, sea-level rise, loss of biodiversity and the degradation of ecosystems.

This volume originates in the hills of Meghalaya at Union Christian College surrounded the Umiam Lake where academicians, scholars and environment experts came together at the International Seminar, *Earth on the Edge: Tackling Global Environmental Crisis*, organised by Union Christian College on February 12 and 13, 2025. This International Seminar emerges as a timely and critical platform for dialogue, collaboration and transformative action. The Seminar brings together a diverse and distinguished group of scientists, policy makers, educators, students and activists who are united by a shared concern for the future of our planet. This gathering is not just an academic exercise; it is a call to conscience, inviting us to re-examine the ways in which human activity has disrupted the natural order and to seek sustainable solutions grounded in research, innovation and ethical responsibility. From global warming and deforestation to biodiversity loss and pollution, the issue we face are interconnected and complex. They require not only technical expertise but also a reimaging of how we, live, consume and relate to the earth. Through the papers presentation, discussions and the deliberations in this seminar, we hope to speak ideas that will ripple outward into classrooms, communities and policies.

I commend all the contributors, organisers and participants for their dedication to this cause. May this book not only raise awareness but also guide a lasting commitment to environment stewardship for the generations to come.

Dr. Rudolf M. Manih
M.Sc., Ph.D, BCS, UB Fellow
Principal
Union Christian College
June, 2025

Preface

Geological records show that the earth has undergone significant changes over billions of years, with natural processes, such as volcanic eruptions, plate tectonics and ice ages reshaping the earth's structure, natural vegetation and climate. The twentyfirst century is, however, witnessing an unprecedented environmental crisis marked by climate change, biodiversity loss, pollution, and unsustainable resource use. The Earth is on the brink, with humanity's actions causing a profound and potentially irreversible impact on natural ecosystems.

Historically, the Earth's ecosystems have been resilient, adapting to shifts in climate, atmospheric composition, and natural resource availability. Yet, as industrialization and urbanization grew, so did the pressures on our environment and resources. Deforestation, overexploitation of natural resources, pollution and the burning of fossil fuels led to rising levels of greenhouse gases, which have contributed to global warming and climate change. In the past decade, however, these challenges have become more acute. Scientific data increasingly show that the Earth's systems are being pushed beyond their natural limits, resulting in more frequent and severe extreme weather events, sea-level rise, loss of biodiversity and the degradation of ecosystems.

According to the data from the National Oceanic and Atmospheric Administration (NOAA), 2016, set a new global temperature benchmark, with monthly anomalies reaching 1.040°C in January, 1.210°C in February, 1.220°C in March, and 1.100°C in April. This dramatic rise in temperatures has been accompanied by increasingly severe climate events. The impacts of climate change are evident across the globe. For instance, devastating wildfires across Australia, Greece, Spain and the United States consume millions of hectares of land destroying habitat and killing animals. Likewise, the rapid melting of Antarctica ice (150 billion ton per year) and Greenland ice (270 billion ton per year) adding to sea level rise, the frequency of catastrophic floods and hurricanes are the impacts of global environmental change.

These escalating environmental crises highlight the urgent and intensifying nature of the crises, affecting ecosystems and human populations alike. The impact on human populations is profound. Vulnerable communities, particularly in developing nations, face the brunt of the crises, often lacking the resources to recover from extreme weather events. Climate-induced migration is on the rise as people are forced to flee from areas that are no longer habitable due to flooding, desertification or heatwaves. In addition, the economic costs of environmental destruction continue to climb, with industries such as agriculture, fishing, and tourism suffering devastating losses.

The crisis has further escalated due to the sheer scale of industrialization, population growth and unsustainable development and consumption patterns. While previous generations had time to adapt to environmental shifts, the rapid changes of the last decade have put us at a crossroads, where urgent action is required to prevent further damage and safeguard the planet's future.

Global ecosystems are deteriorating due to human-induced factors such as deforestation, industrial emissions, overfishing, and urban sprawl. The consequences include rising global temperatures, extreme weather events, diminishing water resources, and compromised air and soil quality. These issues not only threaten biodiversity but also exacerbate poverty, inequality, and public health crises, making environmental sustainability an urgent priority. This moment in history demands reflection, as well as a collective, inclusive effort involving all stakeholders—the vulnerable, the voices of the indigenous peoples, who are often unheard in global environmental discourse, to address the environmental issues that threaten not just the natural world but also human survival. The time to act is now—together, we can pull Earth back from the edge.

This volume originates in the hills of Meghalaya surrounded the Umiam Lake where academicians, scholars and environment experts came together at the International Seminar, *Earth on the Edge: Tackling Global Environmental Crisis*, organised by Union Christian College on February 12 and 13, 2025. The editors in the introductory chapter, *Silent Environmental Warnings*, highlighted the environmental crisis that is plaguing our planet. The editors argue that humanity is in the midst of a biodiversity and climate crisis, and all this at a time when we are beginning to understand that we depend on nature much more than nature depends on us. A nature-positive future needs transformative, game changing, shifts in how we produce, how we consume, how we manage the earth's limited resources and how we govern the environment. This goes beyond conventional conservation. The author concludes that the present generation is at the most important moment in history, while facing its deepest systems change challenge around what is perhaps the most existential of all our relationships: the one with nature, have a chance to act and to mitigate the crisis.

Prof. S.R. Joshi and Nirmala Akoijam in their chapter titled, *Environmental Crisis: Human Greed Overshooting Sustainability*, contend that the earth faces a severe environmental crisis due to unchecked resource consumption and industrial expansion. "Earth on the Edge: Tackling Global Environmental Crisis" highlights how economic growth has caused deforestation, biodiversity loss, pollution, and climate change, pushing the planet towards collapse. Earth Overshoot Day shows resource use surpasses Earth's capacity, leading to severe degradation. Deforestation and fossil fuel reliance worsen climate change and threaten water, food, and biodiversity. Urgent action *via* policy reform, technology, and community engagement is crucial. Governments, businesses, and individuals must adopt sustainable practices. While international cooperation is key to mitigating climate change, immediate intervention to prevent an uninhabitable world is an urgent necessity. Sustainability must be a global priority to put a check on depleting resources and environmental crisis.

Priyangshu Chakraborty and Binesh Safi, in their article, *India's role in Promoting Regional Cooperation for Sustainable Development Goals in South Asian Region,* argue that the region of South Asia covers vast population and disparities in development of which faces critical challenges in achieving the Sustainable Development Goals (SDGs). India, as the prime largest economic stakeholder, plays a significant role in fostering regional cooperation to mitigate collective hurdles. The study explores India's initiatives to promote sustainable development across South Asia through regional frameworks like SAARC, BIMSTEC and bilateral partnerships. The study highlights, the strategic importance of resource utilization, inclusive policies, technological cooperation's to achieve SDGs in South Asian regions.

The fourth chapter, *Disaster Preparedness: Perspectives in West Jaintia Hills, Meghalaya,* authored by Nambha Pohrmen and Tiewri Lyngdoh Nonglait, examine the degree of disaster awareness and readiness among students in schools and colleges. They argue that although most schools cover disaster management in their curricula, few offer specific classes on the subject; instead, they are frequently incorporated into workshops or other subjects. Nonetheless, there are notable deficiencies in catastrophe preparedness. Many institutions lack a dedicated disaster management committee, disaster management maps are rarely visible, and preparedness drills are regularly held. Overcrowding, poor infrastructure, insufficient training, and financial limitations were identified as the main obstacles to successful disaster preparedness.

Prof. R.M. Shangpliang in her chapter, *Environment, Nature and Society: Where do we go from here?,* argue that we are living in a world where Environment is one of the greatest concerns to every society. She contends humans and environment are so closely entangled; it would often be misleading to try to separate them. She held that humans have drastically changed the way we look at 'mother earth'—far from living in harmony with nature we have now reached an age of the triumphant epoch of humans overpowering nature. Faced by these challenges, the earth is today nothing less than "a shrinking island in space". Surely, the most significant issue facing the human species in the present century and beyond is our Ecological Predicament—How can we survive as a species without undermining or degrading the planet Earth and its fabric of life—the very means of our survival? What is the future of rural livelihoods that depend on land and forest resources? Do we recognize the role of tribal women and sustainability? Where do we go from here?

The sixth chapter in the volume, *Knowledge, Awareness and Attitudes Towards Sacred Groves among Students of Shillong College,*

Shillong, Meghalaya, India, by Bennathaniel H. Diengdoh, argued that the management and conservation of sacred groves are not only dependent on the beliefs and perceptions of the communities that have instituted them, but also on those of the larger general populace. Since attitudes and perceptions exert a notable influence on views and practices with regard to conservation, the author argued that greater awareness and a positive attitude towards sacred groves among youth may help to garner publicity and facilitate the promotion of said groves for ecotourism and/or conservation. From his study to ascertain the views, attitudes, perceptions and beliefs of college students of Shillong College, Shillong, Meghalaya regarding sacred groves, the author conclude that a majority of respondents demonstrated a positive attitude towards sacred groves, considering them to be of religious and cultural importance and for biodiversity conservation.

Barnes Mawrie, a noted academician on Khasi Envionmental Ethics, noted in his chapter, *Khasi Traditional Institution to Mitigate the Impact of Climate Change,* that Climate change summits remain an impasse until and unless there is a change of attitude, mentality and a conversion of the human heart. As long as humans do not respect nature, they will continue to exploit it. There needs to be a change of perspective towards our natural environment, to see it as "mother" and not as "matter", to understand our role as "caretakers" and not as "exploiters". Speaking about the Khasis' traditional respect for mother nature, he argues that for the Khasis, nature is perceived as a "subject" to relate with and not as an "object" to be exploited.

S. Hayong, in his chapter, *Re-discovery of the Value of Indigenous Peoples' Ecological Resources: A Model for Environmental Preservation,* argues that indigenous people are considered the first settlers of the land. Nature is not only their sources of living, or their dwelling habitat but it is part and parcel of their life. The author contends that recognizing their value not only help them to conceive and understand about their importance and role to nurture and sustain life rather it also helps them to make significant effort to maintain, preserve and protect them from any forces that may damage them. He calls for a need to rediscover and reimagine the role and importance of these fundamental resources so that they could provide important suggestive measure concerning environmental preservation while at the same time help assist the earth from further ecological crisis.

Ashish Malhotra in his chapter, *Indigenous Practices of Environmental Governance in the North East India: A Brief Review*, argues that the centuries old ecological knowledge of indigenous people is disappearing and if not harnessed now the remaining will be extinct in the near future. Science, with its powerful analytical tools has failed to provide answers and hence the world's disastrous mismanagement of the environment continues to exponentially increase. The author highlighted traditional methods practiced by the indigenous people of North East India which include, traditional rainwater harvesting, traditional water harvesting, bamboo drip irrigation, creation of water tanks for water conservation, preservation of sacred groves, sacred pond, sacred patches of grassland and others have been long practiced by many of the tribes in this part of the country.

The tenth chapter by Bennathaniel H. Diengdoh and S. Jayakumar, *A Comparative Study of Forest Cover and Land Use and Land Cover Change in and around Two Sacred Groves in East Khasi Hills District, Meghalaya from 1990-2020 Using LANDSAT Imagery*, dwell on the issues and changes concerning sacred groves, due to the prevailing land use practices and usage regimes in an area, which can result in their growth or decline. The authors employ remote sensing and analysis of satellite imagery of a given area at predetermined intervals to analyze changes in forest cover within a grove and in land use land cover in the surrounding area. The chapter concludes that the Mawphlang sacred grove, a popular tourist destination with strict protection regimes showed negligible decline in comparison to Swer sacred grove.

Meghna Sarma in her chapter, *Climate Smart Agriculture: An Approach to Transform Agricultural Systems for Food Security,* contends that humanity has to address the fact that almost one billion people go to bed hungry everyday. In addition, the author argues that there is increased competition for land, water, energy, and other inputs into food production. Climate change poses additional challenges to agriculture, particularly in developing countries. She argued that Climate-smart agriculture will result in increased productivity and resilience to environmental pressures, while at the same time reduces greenhouse gas emissions or removes them from the atmosphere, because we cannot ignore the fact that agriculture is itself a large emitter of greenhouse gases.

A joint article by Sushma Gautam, Jyoti Sharma, Sonali Bhambhani, Nandini Bansal, Deepal Sharma, Daksh Massey, Nikhil Dogra and Neha Sharma, *A Sustainable Approach to Circularize Effluent Management in Dairy Industry: A Comprehensive Review,* argue that dairy sector witnesses a significant challenge culminating from lack of stringent effluent management practices; contributing to eco-toxicological manifestations including climate change, eutrophication, acidification, threat to biodiversity and health adversities. They held that complacency in recycling, re-use, reduce policy has increased both demand and supply of depleting resources, thus necessitating strict implementation of CE. Dwindling resources and subsequent generation of enormous volumes of waste water has urged CE interventions. This lucid account corroborates the fact that WWTPs with an operational *circular* model with energy production and resource recovery should be channelized as a component of policy interface. The authors contend that this would necessarily help us inch closer to attainment of Sustainable Development Goals (2&3).

Neha Sharma, Sushma Gautam, Sonali Bhambhani and Jyoti Sharma in their chapter, *Green Pesticides in Agricultural Sustenance: An Attempt to Inch Closer to SDG2*, contend that the unabated use of pesticides has significantly deteriorated soil health and led to eco-toxicological implications. Nevertheless, the efficacious targeted approach dynamics of synthetic/chemical pesticides are reasonable for their implementation in fields, aiming at pest abatement. Agriculture-centric community by virtue of occupational hazards is exposed the most, thereby posing health adversities. The authors argued that vital plant derived ingredients (bio-active compounds) and their bio-efficacy to arrest and manage pest infestation in tomato crop which will align with Sustainable Development Goals 2.

Alfreddie Rani and R.K. Josmee's chapter on *Sustainable Agriculture using Modern Technologies*, argues the ever increasing world's will lead to an increased demand for food, which can only be met by improving crop yields. Therefore, the modernization of the agricultural sector has become crucial. Numerous factors contribute to low crop production, many of which can be mitigated through the integration of advanced agricultural technologies. Unmanned Aerial Vehicles (UAVs) have revolutionized crop health monitoring, a cornerstone of modern agriculture. By harnessing high-resolution imaging and advanced data analytics, drones offer innovative solutions to enhance crop health and optimize agricultural practices. Crop health monitoring plays a critical role in agricultural classification, yield estimation, field management, productivity enhancement, irrigation planning, fertilizer optimisation, sustainable agricultural development, and food security. This chapter provides insights into the effects of environmental and climate changes on crop growth, particularly in Meghalaya, which has a diverse agricultural system. The authors highlight the potential of drone-derived NDVI time series in accurately monitoring crop phenology and health, enhancing productivity, optimizing field management, and promoting sustainable agricultural development across varied climatic conditions.

Banshaikupar Lyngdoh Mawlong
Jonestar L. Nonglait
Editors

Acknowledgements

At the outset, We thank God Almighty for his continued guidance and blessings.

We would like to express our sincere gratitude to all those who contributed to the development and publication of this edited volume. First and foremost, we are deeply thankful to the contributing authors for their insightful chapters, rigorous scholarship, and commitment to excellence. Their diverse perspectives and dedication have enriched this Volume immensely.

We are grateful to our peer reviewers, whose thoughtful feedback and constructive critiques greatly enhanced the quality of each chapter. Their expertise and generosity with time are deeply appreciated.

Our sincere thanks go to the editorial and production team at Concept Publishing Company Pvt. Ltd., New Delhi whose professionalism, patience, and guidance have been instrumental throughout the publishing process. Special appreciation goes to Ashok Mittal Sir for his unwavering support and encouragement.

We offer our sincerest gratitude to Dr. R.M. Manih, Principal of Union Christian College, for his patronage, exemplary guidance, suggestions and inspiration throughout the preparation of the Volume. We also acknowledge the support of our colleagues, especially Dr. J.P. Marak, IQAC Coordinator, for their inspiration, and motivation during the preparation of this book.

Finally, we extend our heartfelt thanks to our families and loved ones for their understanding, patience, and encouragement throughout this journey.

UCC, Shillong

Banshaikupar Lyngdoh Mawlong
Jonestar L. Nonglait
Editors

Contents

Section II

Indigenous Practices and Traditional Knowledge

Section III

Environment and Agricultural Sustainability

List of Contributors

Alfreddie Rani, School of Social Sciences, Central Agriculture University, Umiam, Ri Bhoi District, Meghalaya.
Email: alfreddierani1997@gmail.com

Ashish Malhotra, Dean of Student's Welfare and Head of the Department of Chemistry, Union Christian College, Ri Bhoi District, Meghalaya.
Email: malhotra.ashish116@gmail.com

Banshaikupar Lyngdoh Mawlong, Assistant Professor, Department of Political Science, Union Christian College, Ri Bhoi District, Meghalaya.
Email: ban1787@hotmail.com

Barnes L. Mawrie, Former Principal at Sacred Heart Theological College, Shillong, Meghalaya.
Email: barnesmawrie@gmail.com

Bennathaniel H. Diengdoh, Assistant Professor, Department of Environmental Science, Shillong College, Meghalaya.
Email: diengdoh.benc87@gmail.com

Binesh Safi, Department of Political Science, Assam University, Silchar.
Email: bineshsafi8@gmail.com

Daksh Massey, Department of Bio-sciences and Technology, Maharishi Markandeshwar Engineering College, Maharishi Markandeshwar (Deemed to be University), Haryana.
Email: dakshmassey7@gmail.com

Deepal Sharma, Department of Bio-sciences and Technology, Maharishi Markandeshwar Engineering College, Maharishi Markandeshwar (Deemed to be University), Haryana.
Email: deepal2254sharma@gmail.com

Jonestar L. Nonglait, Department of Geography, Union Christian College, Ri Bhoi District, Meghalaya.
Email: joneslyng@gmail.com

Jyoti Sharma, Department of Bio-sciences and Technology, Maharishi Markandeshwar Engineering College, Maharishi Markandeshwar (Deemed to be University), Haryana.
Email: sushmagautam3255@gmail.com

Meghna Sarma, SMS, Agronomy, Indian Council of Agricultural Research (ICAR), Ri Bhoi District, Meghalaya.
Email: meghnasarma3@gmail.com

Nambha Pohrmen, Department of Environmental Science, Shillong College, Shillong, Meghalaya.
Email: nambhapohrmen3@gmail.com

Nandini Bansal, Department of Bio-sciences and Technology, Maharishi Markandeshwar Engineering College, Maharishi Markandeshwar (Deemed to be University), Haryana.
Email: nandinibansal745@gmail.com

Neha Sharma, Professor and Dean Research (RISE), Chandigarh Group of Colleges, Punjab.
Email: nehamicrobiologist@gmail.com

Nikhil Dogra, Department of Bio-sciences and Technology, Maharishi Markandeshwar Engineering College, Maharishi Markandeshwar (Deemed to be University), Haryana.
Email: dogranikhil907@gmail.com

Nirmala Akoijam, Department of Biotechnology and Bioinformatics, North Eastern Hill University, Shillong, Meghalaya.
Email: naocha16@gmail.com

Priyangshu Chakraborty, Department of Political Science, Assam University, Silchar.
Email: priyangshuchakrabortyaus25@gmail.com

R.K. Josmee, School of Social Sciences, Central Agriculture University, Umiam, Meghalaya.
Email: josmeecpgsas@gmail.com

Rekha M. Shangpliang, Professor, Department of Sociology, North Eastern Hill University, Meghalaya.
Email: socionehu@gmail.com

Sonali Bhambhani, Department of Bio-sciences and Technology, Maharishi Markandeshwar Engineering College, Maharishi Markandeshwar (Deemed to be University), Haryana.
Email: nehamicrobiologist@gmail.com

Sushma Gautam, Department of Bio-sciences and Technology, Maharishi Markandeshwar Engineering College, Maharishi Markandeshwar (Deemed to be University), Haryana.
Email: sushmagautam3255@gmail.com

S. Hayong, Faculty, John Roberts Theological College, East Khasi Hills District, Meghalaya.
Email: shanbhadotsni@gmail.com

S. Jayakumar, Professor, Department of Ecology and Environmental Sciences, Pondicherry University, Puducherry.
Email: s.jkumar1@pondiuni.ac.in

S.R. Joshi, Professor, Department of Biotechnology and Bioinformatics, North Eastern Hill University, Meghalaya.
Email: srjoshi@nehu.ac.in

Tiewri Lyngdoh Nonglait, Assistant Professor, Department of Environmental Science, Shillong College, Shillong, Meghalaya.
Email: tiewrilyngdoh@gmail.com

SECTION I

Environment Crisis and Sustainability

1

Silent Environmental Warnings

Banshaikupar Lyngdoh Mawlong and *Jonestar L. Nonglait*

Introduction

One of Euclid's axioms states, "The whole is greater than the part". We, humans, are just a microscopic part of that whole. Environmental crisis are aplenty in our time. For example, Wildfire choked California from January 7 to 31, 2025 and toxic air fills the Sky of Delhi every winter. A small unknown town in Meghalaya, Byrnihat, has earned the notorious tag of the polluted city in India (IQAir, 2024). COVID-19 is a stark reminder that we are not the kings of the ecosystem, but mere subjects of the complete whole. Our position, on the whole, has been aptly summed up by Chief Seattle when he said, "...the earth is our mother. Whatever befalls the earth befalls the sons of the earth. The earth does not belong to man—man belongs to the earth. The man did not weave the web of life—he is merely a strand in it" (Cited in Mawlong, 2020). Our life is just a link in a chain of innumerable living beings that have succeeded one another on Earth over nearly four billion years. However, as the Great Greek Mathematician, Pythagoras said "Man is the measure of all things", human actions speak to the contrary.

The pages of human history have been diligently dotted with scores of stories of his achievements, glories, and his brilliance. In all those pages little importance, if ever, was given to the provider and supporter of our life, the environment. The environmental discourse was never taken seriously. Left to the periphery of human thoughts and consideration, the environment was exploited to meet man's insatiable needs. However, we cannot deny the truth that human life is impossible outside the earth's natural ecosystem. David Lowenthal in his article "Environmental History: From Genesis to Apocalypse" (1996) has rightly argued that human history cannot and should not be read without a proper reverence to the environment. In doing so, humanity has lost hundreds of years of environmental lessons.

Man's history of its relationship with the environment has seen its fair share of good and bad times. While there has a historical lesson of environmental concern and preservation, events of environmental exploitation are also aplenty. Mankind must remember the truth that the foundation of life, including human life, is the interdependence of all life forms in the ecosystem. In fact, the fabric that supports all life on earth is the sacred ecological balance without which life would be impossible. Mankind has, however, disturbed this fragile, subtle, and sacred balance in the ecosystem. We have changed the face of the world. Despite our vulnerability, we have taken possession of every habitat and conquered swaths of territory like no other species before.

Over the ages, the quest for power has derived to capture the psyche of nature. The fruits of modern in the form of uncontrolled industrialization, unplanned urbanization, consumerism ideals, and overpopulation have wreaked havoc on the earth's limited natural resources. However, like Mahatma Gandhi aptly put, "The world has enough for everyone's need, but not enough for everyone's greed". Our capitalist driven economic model has stripped mother earth bare and naked. Consumerism has driven us blind. And there are no signs of things slowing down. Call for a greener and healthier earth have often been met with mere lip service from statesmen and those in power to act. Mere blanketing of such issues has brought us to where we are today, a global and historic environmental crisis. Scientific data increasingly show that the Earth's systems are being pushed beyond their natural limits, resulting in more frequent and severe extreme weather events, sea-level rise, loss of biodiversity and the degradation of ecosystems.

Are We in an Environmental Crisis?

A crisis, in the context of the environment or ecology, may be defined as a situation where rapid, significant, and often unexpected changes destabilize the survival or well-being of species, populations, or entire ecosystems (Taylor, 2009). The key characteristics that constitute a crisis include:

(i) **Dramatic, Rapid Change:** The situation involves swift and substantial shifts in environmental conditions, such as sudden temperature increases, mass extinctions, or large-scale pollution events. The Sixth Assessment Report (AR6) of the Intergovernmental Panel on Climate Change (IPCC), highlights that human influence has warmed the climate at a rate that is unprecedented at least in the last 2,000 years. It noted that in 2019, atmospheric Carbon Dioxide (CO_2) concentrations were higher than at any time in at least 2 million years, and concentrations of methane and nitrous oxide were higher than at any time in the last 8,00,000 years (Intergovernmental Panel on Climate Change, 2023).

(ii) **Unexpected or Low-Probability Events:** Crises often arise from events that are not easily predicted or that have a low probability of occurring, making them difficult to prepare for or manage. For instance, 21.5 million people are displaced by extreme climate-change related disasters every year (UN Refugee Agency, 2021).

(iii) **Irreversibility:** The changes are difficult or impossible to reverse, leading to long-term or permanent damage to ecosystems or human welfare. The World Wildlife Fund in its 2024 Report held that in the past 50 years (1970–2020), the average size of monitored wildlife populations have shrunk by 73 per cent, which according to the Living Planet Index (LPI), may become self-perpetuating, resulting in substantial, often abrupt and potentially reach a tipping point of irreversible change.

(iv) **Significant Welfare Losses:** The crisis results in substantial harm to the well-being of affected communities, wildlife, or the environment itself. For instance, a 2021 report from Swiss Re Institute, one of the largest providers of insurance to other insurance companies, revealed that climate change could cut the value of the world economy by $23 trillion by 2050 (Swiss Re Institute, 2021).

(v) **Destabilization of Survival:** Changes in abiotic (non-living) or biotic (living) factors disrupt the continued existence or stability of species or populations. The UN Secretary-General António Guterres said that the contemporary environmental siutuation was nothing less than "a code red for humanity. The alarm bells are deafening, and the evidence is irrefutable." He noted that the internationally-agreed threshold of 1.5 degrees above pre-industrial levels of global heating, which could threaten the survival of many species, was "perilously close" (United Nations, 2021).

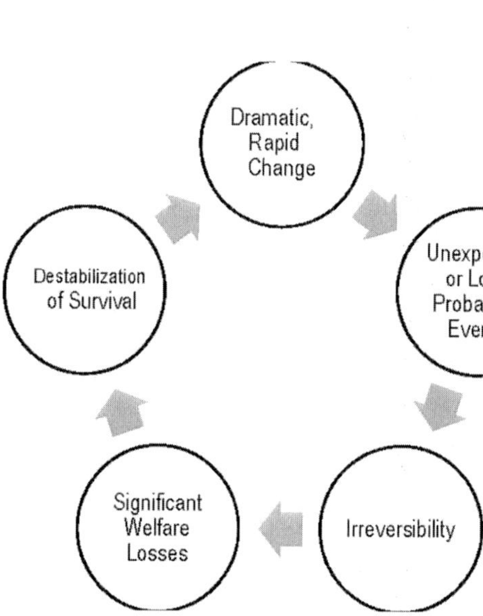

Fig.1.1: Characteristics of Environmental Crisis.

In short, an environmental or ecological crisis is marked by sudden, severe, and often irreversible environmental degradation that threatens survival and welfare, typically caused by a combination of natural events and human activities.

The United Nations and other leading organizations recognize the contemporary environmental siutation, as a "triple planetary crisis", consisting of climate change, pollution, and biodiversity loss (United Nations Framework Convention on Climate Change, 2022). These crises are deeply interconnected and require urgent, coordinated action to prevent irreversible damage to life on Earth. Recent scientific assessments reveal that global temperatures have reached 1.34-1.41°C above pre-industrial levels, with atmospheric CO_2 concentrations hitting record highs and wildlife populations experiencing an average 73 per cent decline since 1970 (World Wildlife Fund, 2024; World Meteorological Organization, 2025). These interconnected environmental changes are generating cascading effects on human health, economic systems, and planetary stability, with economic losses from extreme weather events alone reaching $264 billion in 2022 (Romanello *et al.*, 2023). The convergence of these crisis indicators suggests that the Earth system is approaching critical tipping points that could fundamentally alter the conditions that have supported human civilization for millennia.

A. Accelerating Climate System Changes

Temperature Rise and Warming Projections

The global climate system is experiencing unprecedented warming that is accelerating beyond previous projections. Current global surface temperatures have risen to 1.34-1.41°C above pre-industrial levels, with recent monitoring data showing that 19 of the last 20 months have surpassed the critical 1.5°C threshold established by the Paris Agreement (World Meteorological Organization, 2025). This represents a dramatic acceleration in warming trends, with the 10-year mean heating now reaching 1.14°C above pre-industrial levels (Romanello *et al.*, 2023). Climate models using the Shared Socio-economic Pathway 2-4.5 (SSP2-4.5) scenario project that the 1.5°C threshold will be consistently exceeded by 2024, with global temperatures reaching 2.0°C by 2039-2054 and 2.5°C by 2057-2087, depending on the climate model used (Ma *et al.*, 2022).

The Intergovernmental Panel on Climate Change's Sixth Assessment Report has refined estimates of climate sensitivity, indicating that the equilibrium climate sensitivity now ranges between 2.5°C and 4°C, with a best estimate of 3°C, representing a narrower but higher range compared to previous assessments (Zhou, 2022). The transient climate response to cumulative CO_2 emissions (TCRE) has been assessed at 1.0°C-2.3°C per 1,000 PgC, with a best estimate of 1.65°C, providing more precise understanding of how ongoing emissions will translate into future warming (Zhou, 2022). These refined projections indicate that the climate system is responding more sensitively to greenhouse gas concentrations than previously understood, suggesting that current warming trends will continue to accelerate unless dramatic reductions in emissions are achieved.

Carbon Dioxide Concentrations and Carbon Cycle Disruption

Atmospheric carbon dioxide concentrations have reached at critical levels, with 2023 marking a particularly alarming milestone in the trajectory of greenhouse gas accumulation. The CO_2 growth rate at Mauna Loa reached 3.37 ± 0.11 ppm in 2023, representing an 86 per cent increase above the previous year and establishing a new record high since observations began in 1958 (Ke *et al.*, 2024). This acceleration in atmospheric CO_2 accumulation occurred despite global fossil fuel emissions increasing by only modest amounts, highlighting significant disruptions in natural carbon cycle processes (Ke *et al.*, 2024). The dramatic increase in atmospheric CO_2 concentrations reflects not only continued anthropogenic emissions but also a substantial decline in the Earth's natural carbon absorption capacity.

Perhaps most concerning is the evidence of a major collapse in terrestrial carbon sinks, which have historically moderated the rate of atmospheric CO_2 accumulation. Land CO_2 uptake north of 20°N has declined by approximately half since 2015, dropping to just 1.13 ± 0.24 GtC yr^{-1} in 2023 (Ke *et al.*, 2024). Regional analysis reveals significant carbon losses, including emissions of 0.58 ± 0.10 GtC yr^{-1} in Canada and losses of 0.13 ± 0.12 GtC yr^{-1} in Southeast Asia (Ke *et al.*, 2024). This deterioration of natural carbon sinks represents a dangerous feedback loop in the climate system, where warming reduces the Earth's capacity to absorb anthropogenic CO_2 emissions, thereby accelerating the rate of atmospheric accumulation and further warming.

Methane and Nitrous Oxide: Intensifying Non-CO Climate Forcers

In addition to carbon dioxide, the rapid escalation of methane (CH) and nitrous oxide (NO) emissions is significantly intensifying global warming. Methane, with a global warming potential (GWP) 84-87 times greater than CO over a 20-year period, reached a record 1,923 ppb in 2023—marking the largest annual increase for the third consecutive year (NOAA, 2024; IPCC, 2021). Key anthropogenic sources include fossil fuel operations, livestock, and waste, while feedbacks from thawing permafrost and expanding wetlands are amplifying natural emissions (UNEP & CCAC, 2021; Saunois *et al.*, 2020). Methane now accounts for 20-25 per cent of radiative forcing from long-lived greenhouse gases.

Nitrous oxide, though less abundant, is an even more potent climate forcer, with a GWP of 273 over 100 years and a lifespan exceeding 120 years (IPCC, 2021). Concentrations rose to 336 ppb in 2023, driven mainly by synthetic fertilizer use and manure mismanagement in agriculture (WMO, 2024). Soil warming is further enhancing microbial NO production, and recent studies suggest agricultural emissions may be underestimated by up to 30 per cent (Thompson *et al.*, 2019).

Together, the unchecked rise of CO, CH, and NO reflects a dangerous synergy of anthropogenic and climate-driven feedbacks. These gases collectively undermine natural sink capacities and accelerate warming, reinforcing the urgency for integrated, multi-gas mitigation strategies (Steffen *et al.*, 2018; Forster *et al.*, 2021).

B. Ocean and Marine Environment Transformations

Sea Level Rise Acceleration

Global sea levels are rising at an accelerating rate that significantly exceeds historical baselines and poses increasing threats to coastal communities worldwide. Analysis of 25 years of precision satellite altimeter data from TOPEX/Poseidon, Jason-1, Jason-2, and Jason-3 missions reveals that sea level rise is accelerating at a rate of 0.084 ± 0.025 mm/y^2 (Nerem *et al.*, 2018). This acceleration aligns well with climate model projections and indicates that sea level rise is being driven by climate change rather than natural variability (Nerem *et al.*, 2018). Current projections suggest that if sea levels continue to change at this rate and acceleration, sea level rise by 2100 will reach approximately 65 cm, which represents more than double the amount that would occur if the rate remained constant at the current 3 mm/year baseline (Nerem *et al.*, 2018).

The acceleration in sea level rise reflects the combined effects of thermal expansion of seawater due to warming and increased mass contributions from melting ice sheets and glaciers. This represents a fundamental shift in ocean dynamics that will have profound implications for coastal infrastructure, ecosystems, and human settlements. The precision of satellite altimetry measurements, combined with careful consideration of interannual and decadal variability as well as potential instrument errors, provides high confidence in these acceleration estimates (Nerem *et al.*, 2018). The implications of accelerating sea level rise extend beyond simple flooding risks to include increased coastal erosion, saltwater intrusion into groundwater systems, and permanent inundation of low-lying areas.

Ocean Acidification Trends

The world's oceans are experiencing rapid acidification as they absorb increasing amounts of anthropogenic CO_2, with the Mediterranean Sea serving as a particularly sensitive indicator of global trends. Continuous pH measurements at the Strait of Gibraltar GIFT time series station documented a remarkable decreasing annual trend of -0.0044 ± 0.00006 pH units over nearly three years of monitoring (Flecha *et al.*, 2015). This rate of acidification is consistent with broader patterns observed across the Mediterranean basin, where regional pH decreases range from -0.055 to -0.156 pH units relative to preindustrial levels, depending on the degree of anthropogenic carbon accumulation in specific sub-basins (Flecha *et al.*, 2015).

Ocean acidification represents one of the most pervasive and irreversible consequences of increased atmospheric CO_2 concentrations. The Mediterranean Sea's sensitivity to acidification makes it a crucial indicator system for understanding global ocean chemistry changes (Flecha *et al.*, 2015). The chemical alterations associated with ocean acidification have cascading effects throughout marine ecosystems, affecting shell-forming organisms, coral reefs, and the entire marine food web. The precision of these pH measurements, combined with the consistent downward trends observed across multiple monitoring stations, provides clear evidence that ocean acidification is proceeding at unprecedented rates in the modern era.

C. Biodiversity Crisis and Ecosystem Collapse

Wildlife Population Decline

The global biodiversity crisis has reached catastrophic proportions, with wildlife populations experiencing unprecedented declines across all major taxonomic groups and geographic regions. The World Wildlife Fund's 2024 Living Planet Report documents an average 73 per cent decline in monitored wildlife populations since 1970, representing one of the most comprehensive assessments of global biodiversity trends available (World Wildlife Fund, 2024). This decline rate indicates that wildlife populations are collapsing at a pace that far exceeds natural background extinction rates and suggests that the Earth is experiencing its sixth mass extinction event (World Wildlife

Fund, 2024). The report warns that as the Earth approaches dangerous ecological tipping points, these biodiversity losses pose grave threats not only to wildlife but to the fundamental life support systems upon which humanity depends.

The magnitude of wildlife population decline reflects the combined impacts of habitat destruction, climate change, pollution, overexploitation, and invasive species. 73 per cent average decline masks even more severe losses in certain regions and taxonomic groups, with some populations experiencing near-complete extirpation (World Wildlife Fund, 2024). This biodiversity crisis is particularly concerning because it represents the loss of ecological resilience and the breakdown of ecosystem services that support human societies. The Living Planet Report emphasizes that addressing this crisis will require a massive collective effort over the next five years to tackle the dual climate and nature crises simultaneously (World Wildlife Fund, 2024).

Ecosystem Service Degradation

The decline in biodiversity is accompanied by widespread degradation of ecosystem services that are essential for human well-being and economic stability. Terrestrial ecosystems are experiencing reduced capacity to sequester carbon, as evidenced by the decline in land CO_2 uptake documented in recent carbon budget analyses (Ke *et al.*, 2024). This reduction in ecosystem functionality creates dangerous feedback loops where environmental degradation accelerates climate change, which in turn further degrades ecosystems. The loss of natural carbon sinks represents just one example of how ecosystem service degradation compounds environmental problems and reduces the Earth system's resilience to human impacts.

Marine ecosystems are similarly experiencing degradation of critical services, including fisheries productivity, coastal protection, and climate regulation. Ocean acidification threatens the foundation of marine food webs by affecting shell-forming organisms and coral reefs that provide habitat for countless species (Flecha *et al.*, 2015). The interconnected nature of ecosystem services means that the loss of biodiversity in one system cascades through multiple environmental compartments, creating cumulative impacts that exceed the sum of individual stressors. This ecosystem service degradation represents a fundamental threat to human societies that depend on natural systems for food security, water resources, climate regulation, and disaster risk reduction.

D. Human Health and Economic Consequences

Health Impacts of Environmental Change

The contemporary environmental crisis is generating unprecedented threats to human health through multiple pathways that are becoming increasingly severe and widespread. Climate change is facilitating the expansion of vector-borne infectious diseases, including dengue, malaria, vibriosis, and West Nile virus, as changing temperature and precipitation patterns create more favourable conditions for disease vectors (Romanello *et al.*, 2023). Heat exposure has become a major health threat, with the current 10-year mean heating of 1.14°C above pre-industrial levels already impacting the health and survival of people worldwide (Romanello *et al.*, 2023). The health impacts of environmental change extend beyond direct physiological effects to include mental health consequences, food and water insecurity, and increased vulnerability to extreme weather events.

Air quality degradation associated with fossil fuel combustion and environmental degradation contributes to respiratory and cardiovascular diseases that affect millions of people globally. The Lancet Countdown's comprehensive assessment finds that people's health is increasingly at the mercy of fossil fuel systems that simultaneously drive climate change and create direct health hazards (Romanello *et al.*, 2023). The intersection of environmental and health crises creates compounding vulnerabilities, particularly for marginalized populations who have limited resources to adapt to changing conditions. These health impacts represent both immediate humanitarian concerns and long-term threats to social stability and economic productivity.

Economic Losses and Financial Risks

The economic consequences of environmental degradation have reached unprecedented scales, with financial losses from climate-related events accelerating rapidly across multiple sectors. Economic losses from extreme weather events increased by 23 per cent between the periods 2010-2014 and 2018-2022, reaching $264 billion in 2022 alone (Romanello *et al.*, 2023). Heat exposure has generated substantial economic impacts through reduced labour productivity, with global potential income losses worth $863 billion documented in recent assessments (Romanello *et al.*, 2023). These direct economic losses represent only the tip of the iceberg in terms of the full economic consequences of environmental degradation, which include infrastructure damage, agricultural disruption, and increased healthcare costs.

The economic impacts of environmental change are creating systemic risks to global financial stability through multiple transmission mechanisms. Property values in climate-vulnerable areas face increasing risks of depreciation, while insurance systems struggle to maintain viability in the face of escalating claims from extreme weather events. Supply chain disruptions caused by environmental disasters generate cascading economic effects that extend far beyond the immediate geographic areas affected by environmental damage. The scale of these economic losses is limiting the resources available for climate adaptation and environmental restoration, creating a dangerous cycle where environmental degradation undermines the financial capacity needed to address environmental challenges.

Conclusion

The contemporary environmental crisis represents a convergence of unprecedented threats that are fundamentally altering Earth system functioning and human society's relationship with the natural world. The scientific evidence presented reveals that multiple environmental indicators are simultaneously approaching or exceeding critical thresholds, including global temperature rise reaching 1.34-1.41°C above pre-industrial levels, atmospheric CO_2 concentrations hitting record highs with growth rates 86 per cent above previous years, and wildlife populations declining by an average of 73 per cent since 1970 (World Wildlife Fund, 2024; World Meteorological Organization, 2025; Ke, 2024). These trends are generating cascading effects through accelerating sea level rise, widespread ocean acidification, ecosystem service degradation, and escalating economic losses that reached $264 billion from extreme weather events in 2022 alone (Flecha *et al.*, 2015; Nerem *et al.*, 2018; Romanello *et al.*, 2023).

The interconnected nature of these environmental changes suggests that the Earth system is approaching critical tipping points that could trigger irreversible alterations to planetary conditions. The decline of natural carbon sinks, acceleration of climate warming, and collapse of biodiversity create dangerous feedback loops that amplify environmental degradation and reduce the Earth's resilience to human impacts (Ma *et al.*, 2022 and Ke *et al.*, 2024). The magnitude and pace of these changes exceed the adaptive capacity of many natural and human systems, indicating that the current trajectory of environmental change poses existential threats to global sustainability and human well-being. Addressing this crisis will require immediate and unprecedented action to simultaneously tackle climate change, biodiversity loss, and ecosystem degradation through coordinated global efforts that recognize the interconnected nature of environmental challenges and their fundamental importance to human survival and prosperity.

While previous generations had time to adapt to environmental shifts, the rapid changes of the last decade have put us at a crossroads, where urgent action is required to prevent further damage and safeguard the planet's future. This moment in history demands reflection, as well as a collective, global effort to address the environmental issues that threaten not just the natural world but also human survival. Mankind needs to remind itself that the engine of life is linkage. Everything is linked. Nothing on this earth, however small and microscopic they may be, are insignificant. Life on earth in itself is a miracle. Every living species crawling the surface of the earth to the depth of the seas has a role to play. All life forms are linked to each other. What befalls the earth, befalls all lives on the planet. And human life is no exception to this sacred lineage.

As a keeper of the earth, Man needs to Acknowledge, Respect, and Preserve the natural world. This harmony between humans and nature must become the rule, no longer the exception. We need to restore the ecological balance, not for anyone's sake but for our very own survival. Our cultural ethics and values must start with human concerns for a healthy, sustainable, and safe environment. We need to recreate our society and behaviour in a way that dramatically reduces environmental stress and rebuild a world that is more sustainable. As an earthling, this is our responsibility and duty. This is a matter of moral and cultural and survival ethics. This is an opportunity for us to hit the reset button. This will require concerted government policy and action that will need civil activism and cooperation from the people. The time to act is now—together, we can pull Earth back from the edge.

REFERENCES

Flecha S., Pérez F.F. , García-Lafuente J., Sammartino S., Ríos A.F. and Huertas I.E. (2015). "Trends of pH Decrease in the Mediterranean Sea through High Frequency Observational Data: Indication of Ocean Acidification in the Basin". *Scientific Reports, 5*:16770.

Forster, P., Storelvmo, T., Armour, K. *et al.* (2021). "The Earth's Energy Budget, Climate Feedbacks, and Climate Sensitivity". In *IPCC AR6 WGI: Climate Change 2021: The Physical Science Basis.* IPCC Sixth Assessment Report Working Group 1.

Intergovernmental Panel on Climate Change (IPCC) (2023). *Climate Change 2023: Synthesis Report. Contribution of Working Groups I, II and III to the Sixth Assessment Report of the Intergovernmental Panel on Climate Change.* Geneva, Switzerland: IPCC.

Intergovernmental Panel on Climate Change (IPCC) (2021). *Climate Change 2021: The Physical Science Basis.* Contribution of Working Group I to the Sixth Assessment Report of the Intergovernmental Panel on Climate Change.

IQAir (2024). *2024 World Air Quality Report.* Switzerland: IQAir.

Ke, P., Ciais, P., Sitch, S., Li, W., Bastos, A., Liu, Z., Xu, Y., Gui, X., Bian, J., Goll, D.S., Xi, Y., Li, W., O'Sullivan, M., Goncalves De Souza, J., Friedlingstein, P. and Chevallier, F. (2024). "Low Latency Carbon Budget Analysis Reveals a Large Decline of the Land Carbon Sink in 2023". *Natural Science Review, 11*(12).

Lowenthal, D. (1996). "Empires and Ecologies: Reflections on Environmental History". In T. Griffiths and L. Robin (eds.). *Ecology and Empire: Environmental History of Settler Societies.* Edinburgh: Keele University Press, United Kingdom.

Ma, Nicole, Jonathan, H. Jiang, Kennard, Hou, Yun, Lin, Trung, Vu, Philip, E. Rosen, Yu, Gu, Kristen, A. Fahy (2022). "21st Century Global and Regional Surface Temperature Projections". *Earth and Space Science, 9.*

Mawlong, B.L. (2020). "COVID-19: The Environmental Wake-up Call". *Current World Environment, 15*(3).

Nerem R.S., Beckley, B.D., Fasullo, J.T., Hamlington, B.D., Masters, D. and Mitchum, G.T. (2018). "Climate-change-driven Accelerated sea-level Rise Detected in the Altimeter Era". *Proceedings of the National Academy of Sciences of the United States of America, 115*(9): 2022-2025.

National Oceanic and Atmospheric Administration (NOAA) (2024). *Trends in Atmospheric Methane.* National Oceanic and Atmospheric Administration, Earth System Research Laboratory.

Romanello, M., Napoli, C.D. *et al.* (2023). "The 2023 Report of the Lancet Countdown on Health and Climate Change: The Imperative for a Health-centred Response in a World Facing Irreversible Harms". *Lancet, 402*(10419): 2346-2394.

Saunois, M., Stavert, A.R., Poulter, B., *et al.* (2020). "The Global Methane Budget 2000-2017". *Earth System Science Data, 12*(3): 1561-1623.

Steffen, W., Rockström, J., Richardson, K., *et al.* (2018). "Trajectories of the Earth System in the Anthropocene". *Proceedings of the National Academy of Sciences, 115*(33): 8252-8259.

Swiss Re Institute (2022). *The Economics of Climate Change: No Action Not an Option.* Zurich: Swiss Re Management Ltd., Swiss Re Institute.

Taylor, M. Scott (2009). "Environmental Crises: Past, Present and Future". *The Canadian Journal of Economics, 42*(4): 1240-1275.

Thompson, R.L., Lassaletta, L., Patra, P.K., *et al.* (2019). "Acceleration of Global N_2O Emissions seen from two Decades of Atmospheric Inversion." *Nature Climate Change, 9*(12): 993-998.

United Nations. 2021. "IPCC Report: 'Code Red' for Human Driven Global Heating, Warns UN Chief". https://news.un.org/en/story/2021/08/1097362.

United Nations Environment Programme (UNEP) (2019). *The Global Environmental Outlook (GEO 6): Healthy Planet, Healthy People.* New York: Cambridge University Press.

United Nations Environment Programme (UNEP) & Climate and Clean Air Coalition (CCAC) (2021). *Global Methane Assessment: Benefits and Costs of Mitigating Methane Emissions.* United Nations Environment Programme and Climate & Clean Air Coalition.

United Nations Framework Convention on Climate Change (UNFCC). (2022). "What is the Triple Planetary Crisis?". https://unfccc.int/news/what-is-the-triple-planetary-crisis

UN Refugee Agency (UNHRC) (2021). "Data Reveals Impacts of Climate Emergency on Displacement". https://www.unhcr.org/us/news/stories/data-reveals-impacts-climate-emergency-displacement

World Meteorological Organization (WMO) (2025). *State of the Global Climate 2024.* Geneva: WMO.

World Meteorological Organization (WMO) (2024). "Greenhouse Gas Bulletin: The State of Greenhouse Gases in the Atmosphere Based on Global Observations through 2023". *World Meteorological Organization.*

World Wildlife Fund (WWF) (2024). *Living Planet Report 2024: A System in Peril.* Gland, Switzerland: WWF.

Zhou, T. (2021). "New Physical Science behind Climate Change: What does IPCC AR6 Tell Us?" *The Innovation* (Camb), *2*(4):100-173, November 28, 2021.

2

Environmental Crisis

Human Greed Overshooting Sustainability

Nirmala Akoijam and *S.R. Joshi*

Introduction

Earth, the only habitable planet in the Solar System (so far), once used to be the ultimate source of sustenance and nourishment for all lifeforms. Earth's resources were plenty and powerful and its potential was unfathomable. However, among all the life she harboured, the *Homo sapiens* species emerged to be the most cognitively advanced, evolving from a mere hunter-gatherer to a domesticated agricultural society and eventually a socially highly structured group of organisms. This growth led to revolutions in scientific and technological aspects, which turned Earth into a giant hub of industrial and technological marvels, but at the cost of depleting its natural balance. The theme *"Earth on the Edge: Tackling Global Environmental Crisis"* aligns with the lecture by highlighting how unchecked human greed has driven resource exploitation beyond sustainable limits. As industries and societies prioritize relentless growth over ecological balance, the planet faces irreversible damage, pushing humanity towards an environmental tipping point.

The Environmental Crisis: A Consequence of Human Greed

The Industrial Revolution that took over Europe almost 200 years ago was just the tip of the iceberg. What started as a mere replacement of manual labour with machines for improvement of economic growth escalated to an endless demand for manufactured goods, which in turn led to increased consumption of fossil fuels like coal. Depletion of such environmental resources and the stark absence of stringent regulatory practices led to severe environmental degradation, with major industrial cities experiencing hazardous air pollution from toxic emissions and water contamination from industrial chemicals.

Andrew Simms, a researcher from New Economics Foundation, partnered with Global Footprint Network in 2006, conceptualized the Earth Overshoot Day campaign. This concept hypothesizes the specific day when humanity's annual demand for ecological resources and services surpasses the capacity of Earth to regenerate those resources within the same year. Once this threshold is crossed, an ecological deficit ensues. During this deficit, resources are consumed at a rate that exceeds the Earth's ability to replenish them, and waste is generated at levels that surpass the natural absorptive capacity of ecological systems. The primary factors contributing to this deficit include excessive consumption, inefficient utilization of resources, and unsustainable industrial activities. If the world collectively lived within sustainable limits, Earth Overshoot Day would ideally fall on December 31 or later. However, in recent decades, it has been occurring earlier each year.

Major Environmental Crises Resulting from Overshooting Sustainability

The unchecked use of natural resources and mass deforestation have incurred losses that affect more than the immediate environment. The domino effect is a widely recognized phenomenon in which an environmental alteration initiates a series of subsequent changes, affecting multiple interrelated components within an ecosystem and frequently resulting in substantial and cascading consequences. This section discusses the real-world consequences of human greed pushing planetary boundaries and the subsequent domino effects.

Deforestation and Biodiversity Loss ! Climate Change and Global Warming

Loss of forest cover directly affects the biodiversity that it harbours and disrupts interconnected ecosystems. Deforestation for anthropogenic benefits lead to loss of habitats and impacts microclimates. As plants and animals lose their ecological niches, it triggers a steady decline in their populations eventually leading to species extinction. It is widely understood that trees absorb CO by the process of photosynthesis and store it in their biomass in the form of carbohydrates and cellulose. This enables wide carbon sequestration, which keeps the global temperature in check as CO traps infrared radiation from the Sun. Deforestation, therefore, leads to rise in global temperature and since fewer trees mean less water retention, it eventually leads to desertification and altered weather patterns. A prime example is the Amazon biome encompasses a number of countries like Brazil, Bolivia, Peru, Colombia and Ecuador; around 47 million people depend on this for their livelihood and additionally harbours many species of mammals, birds and amphibians, some of which are still undiscovered. The destruction of the Amazon rainforest for huge-scale farming, illegal and unsustainable logging practises and uncontrolled urbanisation. This has led to loss of approximately 17 per cent of the Amazon forest cover, which is continually being degraded (Living Amazon Report, WWF, 2022).

Climate Change and Global Warming—Water Scarcity and Overuse of Natural Resources

With rise in global temperatures, the rate of evaporation of freshwater sources also increases leading to more droughts. Increased heat also leads to higher water demand in agriculture and human consumption, worsening scarcity. Regions with dry soils and reduced plant cover absorb more solar radiation and heat, fostering high-pressure systems that inhibit rainfall, causing already arid areas to become even drier. Melting glaciers also disrupt water supplies in the long run for areas that depend on snowmelt for drinking water and other uses like irrigation and even hydropower generation. In the remote borders of Pakistan, the melting of the Shisper Glacier led to a glacial lake outburst in May 2022, which destroyed road connectivity and power plants. Farmers in this region who depend on seasonal snowmelt for irrigation of their crops have had to look for alternative measures due to the unpredictability and risks (eroded banks and swept-off lands) associated with glacial lake outbursts (Chandrashekhar, 2022). Erratic rainfall patterns contribute to the depletion of freshwater sources such as rivers and lakes, increasing dependence on groundwater and aquifer reserves. This excessive extraction leads to a continuous decline in the water-table, prompting further overpumping, which in turn triggers land subsidence and facilitates heavy metal contamination of groundwater.

Water Scarcity and Overuse of Natural Resources—Pollution and Waste Crisis

As water shortages become more frequent, communities are

compelled to depend on polluted or overexploited water sources for domestic needs and turn to bottled water for drinking. This not only intensifies plastic waste pollution but also raises the risk of exposure to micro plastics, which can accumulate in the human body and pose potential health risks. Additionally, the scarcity of clean water often necessitates the reuse of wastewater, requiring extensive chemical treatment to remove contaminants. However, this process can generate harmful by-products, including toxic sludge and chemical effluents, which, if not properly managed, seep into soil and water bodies, further degrading environmental quality. In regions with inadequate water infrastructure, untreated or partially treated wastewater is often discharged into rivers and lakes, threatening aquatic ecosystems and exacerbating public health risks.

Pollution and Waste Crisis—Deforestation and Biodiversity Loss (Cycle Repeats)

Air pollution, primarily caused by the burning of fossil fuels, leads to climate change, which alters habitats and disrupts ecological balance. Land pollution, including improper waste disposal and deforestation, depletes soil nutrients and reduces the availability of suitable habitats for wildlife. Water pollution, caused by industrial discharge, oil spills, and excessive plastic waste, contaminates aquatic environments, leading to the loss of marine species. As ecosystems deteriorate, many species struggle to adapt to the rapidly changing conditions, ultimately resulting in population declines and, in severe cases, extinction. The loss of biodiversity further weakens ecosystems, making them less resilient to environmental changes and human-induced pressures. Soil pollution, particularly from industrial waste and chemical contamination, seep into the soil, rendering it infertile and unfit for cultivation. As a result, agricultural expansion becomes necessary, leading to deforestation as more land is cleared for farming. This destruction of forests reduces carbon sequestration, further exacerbating climate change. The degradation of soil quality also affects local communities, as they struggle to produce sufficient crops, leading to food insecurity. Contaminated soil can introduce harmful chemicals into the food chain, affecting human health. Thus, soil pollution not only threatens environmental sustainability but also has long-term consequences for both biodiversity and human populations.

The excessive accumulation of plastic waste in oceans has become a critical environmental issue, disrupting marine biodiversity and threatening entire ecosystems. Plastic debris, often mistaken for food by marine animals, causes internal injuries and death in species such as sea turtles, seabirds, and fish. Additionally, micro plastics—tiny plastic particles resulting from the breakdown of larger plastic waste—enter marine food chains, ultimately reaching humans through seafood consumption. Moreover, plastic pollution damages coral reefs, which serve as essential habitats for marine life. Plastic accumulation on the ocean floor alters the physical structure of marine ecosystems, affecting species that rely on stable environments. As marine biodiversity declines, the balance of oceanic ecosystems is disrupted, leading to cascading effects on global fisheries and food security.

Over time, these negative impacts compound, creating a vicious cycle in which human activities further degrade the environment, leading to even more severe consequences for both nature and society. Without intervention, this escalating cycle of destruction threatens to reach a tipping point where restoration becomes nearly impossible, leaving future generations to inherit an increasingly uninhabitable planet.

The Path to Sustainability: Solutions and Future Directions

Addressing environmental degradation requires a multifaceted approach that integrates policy reforms, technological innovations, and community-driven initiatives. Governments play a crucial role in enforcing strict environmental regulations, such as limiting industrial emissions, promoting sustainable land use, and banning single-use plastics. Incentivizing businesses to adopt eco-friendly practices, such as using biodegradable packaging and investing in renewable energy, can significantly reduce pollution. Additionally, the development of green technologies, including carbon capture systems, waste-to-energy processes, and biodegradable materials, offers promising solutions for minimizing environmental damage.

Public awareness and community participation are equally essential in the journey towards sustainability. Educational campaigns can encourage individuals to adopt environmentally responsible behaviours, such as reducing waste, conserving energy, and supporting sustainable products. Grassroots movements and non-governmental organizations (NGOs) also play a vital role in advocating for policy changes and implementing conservation projects, such as afforestation and marine clean-up initiatives. Moreover, adopting circular economy principles—where resources are reused, recycled, and repurposed instead of discarded—can help reduce waste generation and dependence on finite natural resources.

International cooperation is critical in tackling global environmental challenges. Climate agreements, such as the Paris Agreement, emphasize the need for collective action in reducing greenhouse gas emissions. Countries must work together to share sustainable technologies, provide financial support to developing nations, and establish global policies that protect ecosystems. Moving forward, the path to sustainability requires a balance between economic development and environmental preservation. By fostering a culture of sustainability, prioritizing eco-friendly innovations, and enforcing responsible policies, societies can mitigate environmental degradation and ensure a healthier planet for future generations.

Conclusion: A Call for Immediate Action

The urgency to address environmental crises has never been greater. Pollution, deforestation, and climate change continue to degrade ecosystems, pushing many species towards extinction and threatening human livelihoods. If immediate action is not taken, the damage may become irreversible, leaving future generations with an uninhabitable planet. Every effort, whether reducing waste, conserving energy, or supporting green initiatives, contributes to a larger global solution. International cooperation and innovation are essential in reversing environmental degradation and building a sustainable future. The time to act is now—if we continue to exceed sustainability limits, the consequences will be devastating and irreversible. By working together, we can restore balance to our planet and ensure a healthier, more resilient world for generations to come.

REFERENCES

Chandrashekhar, V. (2022). *As Himalayan Glaciers Melt, a Water Crisis Looms in South Asia*. Yale Environment 360. https://e360.yale.edu/features/himalayas-glaciers-climate-change

Vergara, A., Arias, M., Gachet, B., Naranjo, L.G., Román, L., Surkin, J. and Tamayo, V. (2022). *Living Amazon Report 2022*. World Wildlife Fund.

3

India's Role in Promoting Regional Cooperation for Sustainable Development Goals in South Asian Region

Priyangshu Chakraborty and *Binesh Safi*

Introduction

In the terminology of the importance of the sustainable world, South Asia holds one of the largest geographical entities with a diverse region of the Asian continent representing unique characteristics and significant status of regional and sub-regional dynamics. The region is optimized with the presence of mainly eight countries, namely Afghanistan, Bangladesh, Bhutan, India, Maldives, Nepal, Pakistan and Sri Lanka relying within the surface of Himalayan Mountains to the Indian Ocean and reflects richness, strategi and strategic alignment with the rest of the world. The region is considered as the most density populated region of the world where 1,830 million people live with an area of 52,20,587 sq. km (Sarkar *et al.*, 2013). Maldives and Bangladesh are the two most densely populated countries in South Asia. This region is considered as one of the world's fastest growing regions in terms of GDP and population growth and even home of roughly a quarter of the world's population, witnessing several correspondence dimensions and possibilities of development since the last two decades. The region also contains world's largest working age population, one-fourth of the world's middle-class consumers, the largest number of poor and undernourished in the world, and pockets of deep poverty, rising inequality, energy shortages and high rates of malnutrition. Although, the region currently experiencing multiple severe challenges claims with destitution and food insecurity to infrastructure and investments, and environmental degradation which often boost with the presence of socio-economic inequalities and unemployment in the region but it is true that South Asia is a hub of unexplored potentialities and had arranged to emerge as captivating pinpoint of discussion in strategic world forums for so long and able to progress in human capital index and holding improvement in arenas of regional stability. The region has even made ultimate progress through trade, economic integration, management of shared waters resources ensuring regional cooperation, which often leads to focus on achieving Sustainable Development Goals (SDGs) in the region. The motive of accelerating towards the goals ensures the need of peace and stability to be able to concentrate on their economic development and achieve the UN's Sustainable Development Goals (Dutt, 2023) within the spheres of international growth and resemblance of legitimate elements for the existence of the human generation both present and future. The Sustainable Development Goals (SDGs) adopted by the United Nations in 2015, focus on implementing its goals by 2030. Its Agenda for Sustainable Development says that "There can be no sustainable development without peace and no peace without sustainable development" (United Nations, 2015). The essence of development determines with a milestone in its engagement of development agenda was determined as

Millennium Development Goals (MDGs) where the region has progressed over the past few years but often remains a big challenge to achieve its targets, for the presence of its political, geographical, security, socio-economic disagreements and inadequate solutions of existing problems. Moreover, the importance of globalization and regional integration or cooperation are often considered as one of the important components in the relevance of global diplomacy and mitigate with asymmetrical challenges to growth with multiple actors. It is driven by a shared goal among nations to achieve peace and economic prosperity through cooperation (Kim, 2024) which reflects the strategic dimensions of development and is explored to achieve sustainability and tackle environmental problems in order to uplift significant development of human needs. India has undergone a remarkable transition in its dynamics of engagement for regional cooperation with evolving geopolitics of world order. The focus represents strategic partnership with neighbouring countries and exploring the potential of optimistic opportunities for achieving sustainable development in the region and well-being service. The idea of sustainability is not alien to India and hence the commitment towards SDGs is central to the country's own development trajectory as sustainability is historically ingrained in India's ethos (Chaturvedi, 2021). The state is considered as an emerging global power and occupies a prominent role amongst South Asian countries, consisting of more than half of the geographical area, population, presence of strong economic, technical and strategic benefits. The nation tries to retain its transformation by reincarnation of its foreign policy and developmental orientations, attempting for collective growth and reliance in South Asian regions to the region's autonomy. India's economic growth, political stability and willingness to share its technological development with neighbours will encourage and sustain the entire region (Chattopadhyay, 2008). The sustainable development goals are considered as essential segments of achievements with global assurance by 2030 to which India has aligned its developmental priorities in both internal (within the state focus on ensuring strong capabilities and productive orientations of networks primarily by NITI Aayog and other government agencies) and external (beyond the borders by engaging with regional cooperation and global initiatives) spheres of assertive significance. To grab the initiatives in order to achieve sustainable development goals and maintain its engagement, the nation has fostered its regional cooperation and strategic dynamics with neighbouring countries reflecting its wholesome perspective of self-respect and self-reliance, expressing *Vasudhaiva Kutumbakam* (an ancient Indian phrase meaning 'the world is one family') and motive of attention to resolve challenges in region. Its greater scope of developmental momentum brings shape to high-edge manufacture, healthcare services, poverty eradication, environmental conservation, technology advancement, trade and economic integration, good governance and sustainability alignment with the regional actors for productive growth building trust and stability for a better future in the region. In this context, India's motive to assess achieving SDGs since the inception of the agenda has been recognized as a holistic approach and focused on better opportunities in the region of South Asia, holds the cooperative notion to regional optimization and showcasing sustainable assessment regardless of the presence of border disputes, inequalities of several segments and other diverse configurations.

Objectives of the Study

The present study tries to understand India's role in promoting regional cooperation for achieving Sustainable Development Goals (SDGs) in South Asia and to explore India's engagement with regional sustainability initiatives in South Asia.

Methodology

For the purpose of the study, the descriptive method has been adopted and collected data is primarily based from secondary sources such as books, journals, magazines, reports published by government and think tanks.

The Sustainable Development Goals and South Asia

Climate change has brought a great concern for the growing problem of the environment and its prospects of resources. With the atmosphere becoming increasingly unstable, rising ocean levels, and more frequent extreme weather events, time to intervene is running out to prevent catastrophic consequences (Li *et al.*, 2021). To reduce the vulnerability of such environmental degradation, the countries have ensured to integrate with several programmes and governance mechanisms for sustainability, to which the United Nations is often considered to play a vital role for its engagement with member states and arrange a global platform for discussion and legitimate results. The SDGs are a crucial and renewed development framework for countries, including South Asian regions to flourish with its implementation because of the presence of similar socio-economic characteristics and challenges in the parameters of engagement. It follows with a low level of development, lack of basic amenities, experiencing of poverty and inequalities, which leads to focus on adoption sustainability targets to secure well-being of the people and nations as well. Additionally, the region is distinct in terms of which Maldives recognised as Small Island Developing State (SIDS), Bangladesh as a least developing country and Afghanistan, Bhutan and Nepal being also landlocked least developing countries considering three common dimensions, i.e. population, GDP and GDP per capita, as South Asia can be grouped into various clusters of countries. Since, the inception of Sustainable Development Goals, the implementation of its elements was considered as one of the prime strategic motives of the member countries of the United Nations, agreed towards a world of sustainability and urge to maintain the developmental agenda. The history of SDGs has a long significance as the term was used in 1987, following this with 'The Earth Summit,1992' where global leaders accept to implement Agenda 21 and then it keeps emerging with changing complex phenomena and finally, the 17 Sustainable Development Goals adopted by all UN member states in 2015 as part of the 2030 Agenda for sustainable development, primarily consists of set of 169 targets and 232 indicators with a time frame of 15 years initiatives, i.e. 2016-2030 for countries to achieve the targets and implementation of respective SDGs. The step was a landmark initiative in the accumulation of the transformative agenda, promoting national policies of states, international partnerships, ensuring participation of multiple private stakeholders to engage and enhance the success of the implementation of Sustainable Development Goals (See Table 3.1).

Table 3.1: Sustainable Development Goals (SDGs)

Numbers	Goals
1.	No Poverty
2.	Zero Hunger
3.	Good Health and Well-being
4.	Quality Education
5.	Gender Equality
6.	Clean Water and Sanitation
7.	Affordable and Clean Energy
8.	Decent Work and Economic Growth
9.	Industry, Innovation and Infrastructure
10.	Reduced Inequalities

(Contd.)

Numbers	Goals
11.	Sustainable Cities and Communities
12.	Responsible Consumption and Production
13.	Climate Action
14.	Life on Water
15.	Life on Land
16.	Peace, Justice and Strong Institutions
17.	Partnerships for the Goals

These ensure the productivity of countries alignment to achieve the goals such as poverty (SDG 1) as an important element required to achieve access to basic necessities, (SDG 4) for quality education and experience, (SDG 9 and 12) for environmental conservation and similarly all other goals with their rationale worthiness. However, the implementation process lens with several challenges in order to achieve its targets and maintain stability, with the deadline of last steps to ensure the productivity of the sustainability by 2030 countries arrange their productive presence to effective policy implementation, governance mechanisms, formulation of cooperation and monitoring the subscription of environmental conservation globally. The Sustainable Development Goals Report of 2024, reveals that progress has ground to a half or been reversed across multiple fronts, despite a reaffirmed pledge (The Sustainable Development Goals Report, 2024: 3). The cumulative impact of the environmental crisis is to prioritize more for which the engagement to achieve sustainability claims as a proactive notion for almost all the countries and engage in a common platform to introduce preservation as utmost importance creating prominent sections to fulfil developmental ends.

India's Role in Promoting Regional Cooperation for Achieving Sustainable Development Goals (SDGs) in South Asia

India's Institutional Engagements in South Asia

The regional priorities for SDG implementation in South Asia can be addressed by making use of connectivities as regional instruments, as the five key connections need to be considered:

(i) Trade Connectivity
(ii) Investment Connectivity
(iii) Transport Connectivity
(iv) Institutional Connectivity
(v) Knowledge Connectivity (Rahman, 2018).

However, India has aligned towards the SDGs as a central initiative defining scope for policies and actions and their integration to promote regional cooperation with the neighbouring countries for sustainability. The following discusses the strategic assessment of the nation's engagement:

South Asian Association for Regional Cooperation (SAARC)

As a founding member of SAARC, India being the largest country and economy in the South Asian region, it ensures sub-regional cooperation process for development in several indicators of SDGs and achieving regional projects related to infrastructure, economic and trade, socio-cultural development and technological dimensions. The contribution of India with SAARC deals with the paradox of India's ethos and farsighted engagements with neighbouring nations for development and optimization of security. The presence of several segments within the association, as SAARC Development Fund, initiated by 2010 with financial mechanisms to promote regional cooperation with mainly three elements, namely social, economic and infrastructural improvements. India reflects its participation with member countries to reach SDGs with effective policy formulation and availability of resources to eradicate poverty, improvement in infrastructure and trade initiatives, cross-border investments and conservation management. The nation has boosted its initiatives in encouraging women empowerment and regional stability, keeping SDGs an early factor for achievement. India hosts SAARC Disaster Management Centre and plays its presence of disaster management, its preparedness and action of response effectively. As a larger player in the region with a strong position of resources, the nation promotes its regional cooperation in humanitarian intervention. Its emergence as a top-ranking Humanitarian Assistant and Disaster Relief (HARD) provider has set special trends on how to provide relief to affected nations (Singh, 2022: 95). The instances of operations in Bangladesh after the cyclone Sidr (2007), Operation Insaniyat supported Rohingya populations during September 2017, Operation Devi Shakti helped the evacuation of Indians and Afghans after the 2021 Taliban control over Afghanistan (Singh, 2022: 96). Operation Karuna and Sadhav in Myanmar after devastated by cyclone Mocha and emergency food assistance both in 2023 and 2024 respectively. India has been the first responder and massive endeavour after the Nepal earthquake in 2015. Under operation "Maitri" there was a massive deployment of the Indian Army, Air force, and specialized teams to rescue those trapped under the debris and even several non-governmental organizations were also involved in relief assistance, the total support from the Indian government, state governments, and Indian NGOs combined amounted to INR 400 crore (US$ 61m) (Upadhyay, 2021: 170).

One of the most remarkable aspects of India's regional policy in recent years has been the launch of the so-called "South Asia Satellite" (GSAT-9) in May 2017, a geosynchronous communications satellite dedicated to offering services to the region, such as telecommunications and broadcasting applications, including telemedicine, tele-education and disaster management support (Modi, 2017). The focus on South Asia Satellite has emerged as a significant feature of international cooperation aligned with space exploration and India's vigorous initiatives reflect its potential instrument for regional mechanisms and provision of governance of space activities. The nation has even ensured bilateral ties with neighbouring countries on space diplomacy arenas as significantly with Bangladesh signed a paper for cooperation in space research. India has close networks to Nepal's telecommunication and broadcasting services. The nation has agreed to help in launching Bangladesh's first communication satellite 'Bangabandhu 1' (Hassan, 2001) and work closely with Bangladesh (SPARRSO) for telecommunication, remote sensing, broadcasting and other space related areas (Gupta & Raju, 2016). Additionally, The SAARC Development Fund, SAARC Agriculture Centre, SAARC Food Bank and SAARC Forestry Centre hold the potential for improving institutional connectivity in South Asia (Khatun & Saadat, 2020) to which India engage its initiatives and manage its cooperation for the growth of regional sustainability. The South Asian Free Trade Area (SAFTA) has been considered an effective arm for India's growth in regional cooperation to achieve SDGs in the region itself, as India unilaterally provides duty free access to maximum items to the least developing countries in the region with zero tariffs. These initiatives ensure India's engagement for sustainability growth smoothly in the region with investment in infrastructure and capacity-building mechanisms. India is engaged with its schedule commitment of SAARC agreement on Trade in Services (SATIS) enforced in 2012, provides for expanding intra-regional investments, trade liberalisation in service sectors (Reddy, 2020). The regional cooperation with

SAARC involves its engagement with South Asian Regional Standards Organisation (SARSO) since its inception in 2011 with the amalgamation of technical configuration, conformity to assessment procedures effectively across the region of South Asia to facilitate the regional technical barriers to trade (TBTs) for quality advancement and safety of goods and services on multiple sectors, determining the SDGs in the region. The nation has also articulated its potential to South Asia Cooperative Environment Programme (SACEP) to often promote regional cooperation on environmental issues and ensure preservation of natural resources, to which India aligns with several programmes and initiatives for sustainability and responsible governance.

Bay of Bengal Initiative for Multi-sectoral Technical and Economic Cooperation (BIMSTEC)

India ensures its regional cooperation with strategic adaptation of collaborative instruments and restoring its traditional links with neighbouring countries. The engagement of India's role in regional cooperation with determination to achieve a sustainability lens with BIMSTEC (Bay of Bengal Initiative for Multi Sectoral Technical and Economic Cooperation) has emerged as a gradual process. This initiative has geopolitical testimony in terms of strategic planning to India's Foreign Policy but essentially, to focus on the engagement to sustainable development goals it interlinked with the advantage of the cooperation as it connects South Asia with Southeast Asia to which India aligns with several programmes and integration. Generally, in South Asia, five nations are categorised as rim whereas two (Bhutan and Nepal) are landlocked nations, still depending on the Bay of Bengal for maritime trade. Today BIMSTEC has acquired a more solid footing as its economic fundamentals are strong with dynamism in growth and ranks among the fastest growing economies in the world with a rate of 7.4 per cent (Ginwal *et al.*, 2022: 71). Its contribution to BIMSTEC's economic cooperation deals with liberalization of trade mainly through the implementation of the BIMSTEC Comprehensive Free Trade Agreement (BIMSTEC FTA) ensuring seamless trade with the member countries with reduction of tariffs led the engagement of regional cooperation in a next level dignity, this often creates wholesome attention towards the achievement of SDGs as a greater scope of development. India has understood the untapped potential of the BIMSTEC's trade connectivity and visualized the BIMSTEC's FTA as a vital alternative to enhance economic integration of the region (Kumar, 2020). BIMSTEC significantly offers India with a variety of opportunities towards the development of SDGs, as being the major player the nation integrates with national and global initiatives for promotion of sustainability ensuring productive networks to infrastructure, investments, trade and commerce, technical and strategic dynamics respectively. The region of BIMSTEC experiences huge economic losses due to disasters resulting from changes in climate conditions, such as strong winds, storm surges, tsunamis and floods (Khatun, 2019: 4). India has launched the BIMSTEC energy centre to renewable energy for assurance of regional sustainability and responsible governance, enhancing the commitment in order to achieve SDGs. Its total CO_2 emission in 2017 was 22454.77 megatons, the highest in the BIMSTEC region, of which 47 per cent came from the power and energy sector, followed by agriculture, transport and construction sectors and also been the world's fourth-largest 31 emitter of carbon dioxide (Khatun, 2019: 6). The country emerged as the largest economic and energy consumer in the region, effectively focusing on regional energy consumption and promoting to engage with broader aspects of climate security, instances like India-Bangladesh-Myanmar gas pipeline project, cross-border electricity projects in Nepal, Bhutan, Bangladesh and several other developmental initiatives. The BIMSTEC energy centre ensures India's engagement to regional cooperation with optimization of SDGs as the objectives determine coordination and strengthening the energy sector, intra-BIMSTEC cooperation and capacity building and including reforms, regulations and energy efficiency. In addition, India collaborates with BIMSTEC countries on strategic coastal security within the Bay of Bengal, considered as geostrategic significance in the contemporary world. With the presence of dominant naval power, India engaged its maritime security enhancing regional security, capacity development and joint account services. India has promoted naval diplomacy with the increasing geo-economic connectedness in the dynamics of SAGAR (Security and Growth for All in the Region)—essentially meaning, "All boats (regional countries) rise with the rising tide (together)" (Khurana, 2018). The engagement with the member countries to enhance the SDGs in South Asian regions, India manifest its special relations with neighbouring states by taking several steps of cooperation like India-Bangladesh Protocol on Inland Water Transport and Trade (PIWTT), MILAN—a biennial multilateral naval exercise which involves BIMSTEC member countries and Indo-Pacific navies, several joint patrols and surveillance of exclusive economic zones with neighbouring countries, it has also engaged to build naval training to BIMSTEC members within the land of India and even established BIMSTEC coastal security forum to improve intelligence sharing and coordinated response for maritime threats. As, the integration would mitigate the non-traditional threats such as trafficking of narcotics, weapons and people, illegal exploitation of natural resources, piracy and sea-based terrorism, damage caused by natural disasters and issues of environmental degradation and climatological changes (BIMSTEC Coastal Security Workshop, 2019). Moreover, India, as the largest and most technologically advanced country within the BIMSTEC region ensures its capabilities to serve regional cooperation and enhancing the commitment of agricultural research, practices, with achieving United Nation's SDGs. India has been the major exporter of grains in the region helping stabilize food supplies and prices, the Target Public Distribution System (TPDS) ensures the productivity of food security initiatives, financial support to farmers have encouraged the advancement in the region and also bringing close ties with neighbouring countries to adopt and ensure oriented growth for productive spheres of sustainability in the region.

Bangladesh-Bhutan-India-Nepal (BBIN)

The Bangladesh-Bhutan-India-Nepal (BBIN) initiative is a sub-regional framework of countries in the eastern geography of South Asia, which has a rich social, cultural, lifestyle and economic exchange (Sinha, 2024). India plays an intense role in encouraging economic integration, energy enhancement, environmental preservation and social development within BBIN to advance United Nation's SDGs in a way that is more effective. The idea of South Asia Growth Quadrangle (SAGQ) gives a path breaking steps to fructify the BBIN pact to facilitate border movement of passengers and cargo (Sinha, 2024). India has invested in industrial corridors and economic zones, ensures to bring fruitful results in free trade agreements with neighbouring countries to bridge the gap of growth and advancement by reducing barriers, assure productive growth in disaster management and biodiversity conservation and similarly developed regional electricity grids, hydropower projects in neighbouring countries (Nepal and Bhutan) and Bangladesh-India Friendship Power plant enhances energy cooperation which eventually promotes capabilities of sustainability in the region. The BBIN-MVA provides an opportunity to member countries to facilitate seamless cross-border transportation and the opportunity for the creation of regional supply chains across the sub-region through backward and forward linkages. This is also likely to result in an increase in trade volumes between member nations and can help in transforming the corridor into a developmental corridor (Taneja *et al.*, 2024). The BBIN Motor Vehicles Agreement (MVA) was signed in 2015 with a similar goal in relation towards the engagement in the region focusing on cross-border transportation. This agreement has the potential of providing economic opportunities to its member countries in the form of making their trade more competitive through a substantial reduction in the cost of transportation, in particular for the landlocked countries like Bhutan and Nepal (Kumar, 2017). In the context of India, the BBIN MVA ensures as helpful for its commitment to promoting regional connectivity and accelerating in creating transport corridors, focusing on north-eastern part of the state. The anticipation is that the main cities of the north-east—such as Guwahati, Shillong and Agartala—would turn into economic hubs, sending ripple effects to other parts of the region (Chaudhary *et al.*, 2015:13). The two important road routes under BBIN MVA for transportation through the North-east India are mostly likely to be, firstly, Kolkata-Dhaka-Sylhet-Tamabil-Dawki-Shillong-Guwahati and secondly, Agartala-Akhaura-Dhaka-Kolkata (Figure 3.1).

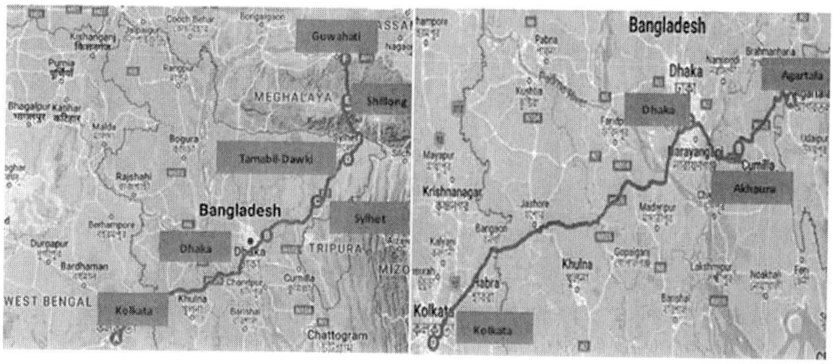

Fig. 3.1: The two important roads of BBIN MVA.

Source: Mapped by ICRIER.

Although the present scenario stating the route is yet to explore but the initiatives are determined with sustainability in the region, since the past few years, the North-east India has explored with a variety of opportunities and potential growth in its indigenous industries reflecting rise in export and importance in the lens of geostrategic geographical locations. The hydropower, as renewable energy and low carbon source, plays its significant action in achieving sustainability in regional cooperation for India's alignment as the region of BBIN blessed with natural resources and such geographical features makes suitable for the development of such initiatives. The vast renewable hydropower potential of Bhutan and Nepal, development of hydropower and cross-border interconnections have the potential for the four countries to improve energy supplies, promote navigation, alleviate floods and encourage connectivity, and add significantly to socio-economic development and even improved regional cooperation can benefit the millions of people in the region who live in extreme poverty and whose lives can be improved with initiatives that support economic growth (Dhakal *et al.*, 2019). As the South Asia region, where a large portion of people living in this region face chronic energy shortages, erratic and unreliable water supply of questionable quality, and rising impacts of climate-change disasters such as floods and droughts and hence, the very real and imminent need for cooperation (Scott *et al.*, 2019). India has engaged in hydropower cooperation with neighbouring countries for sustainability achievement as with Bhutan for Chhukha Hydropower Project (336MW), Tata Hydropower Project (1,020 MW) and Panatsangchhu-I (1,200MW) with financial and technical collaboration (Chakradeo, 2019), with Nepal as Arun-III Hydropower Project (900MW) (India News Network, 2024) and Upper Karnali Hydropower Project in Achham and Dailekh districts of (900MW) (Prasain, 2024). In addition to it, India's vaccine diplomacy has been at the centre of its 'Neighbouring First' policy during COVID-19 pandemic (Singh, 2022), it often gets explored as with the North-eastern region of having several pharmaceutical units and significant supply chain, ensures large proportion of exports to destined neighbouring countries.

India's Leadership in Achieving Key Sustainable Development Goals (SDGs) in South Asia

The SDGs serve as a comprehensive blueprint aimed at achieving a better and more sustainable future for all. The goals call for action on addressing global challenges such as poverty, inequality, climate change, environmental degradation, peace and justice (Sustainable Development Goals, 2024). In South Asia, India being largest provider has launched multiple domestic and regional initiatives to enhance with the SDGs and ensure benefit of the region. These initiatives are arranged with the indicators of SDGs, discussed below:

(*a*) *No Poverty and hunger* (*SDG 1 and 2*): India's leadership in eradicating poverty beyond borders, which benefits South Asia with regional cooperation initiatives, India, manages to contribute in SAARC Development Fund focusing on poverty alleviation in member states and ensures humanitarian aids to manage poverty driven crises in the region. These include importantly cooperation in a sub-regional food security system including through the SAARC food bank and through policy changes that bring about a unified market for essential staples and other food items to prevent temporary shortages and stabilize prices of food for the sub-region's population (Chaturvedi, 2021: 39). It ensures to export food supply to neighbouring countries enrich in varieties of products and sustainability, the South Asia Food and Nutrition Security Initiative is considered to be as a strong pillar for its successful operation.

(*b*) *Health and Education* (*SDG 3 and 4*): India plays a significant role in promoting regional integration with ensuring SDGs focusing on healthcare across South Asia, as the nation holds the largest pharmaceutical industries, professionals, advanced treatment and even provides medical assistance, arranges towards the development of the region. The Vaccine Maitri or Diplomacy has been ensured by the nation towards neighbouring countries during the COVID-19 pandemic. The establishment of the SAARC COVID-19 Emergency Fund with contributions from all SAARC countries, cooperation on COVID-19 has opened the door for possibilities of cooperation in other areas of health and the SDGs in general (Chaturvedi, 2021: 40). Again, India also enhances educational standards in South Asia through the nation's capabilities of research, training and educational curriculum. With the prestigious universities and educational institutions in the region, it offers scholarships to foreign students to study in India and even support native students abroad. The SAARC Education Cooperation and Skill India Programme Expansion with neighbouring countries have been engaged with people to people exchange, ensuring growth of sustainability.

(*c*) *Cooperation in Clean Energy, Economic Growth and Industry, Innovation and Infrastructure* (*SDGs 7, 8 and 9*): India has ensured its potential instrument in clean energy and climate security, which highlights the leading position of regional cooperation. The ISA is a huge stride forward for India as the organisation expands its presence in the globalised globe and provides answers to two crucial issues facing the modern world, namely, climate change and energy production (Hussain and Khan, 2023: 2130). It has enhanced cross-border connectivity and electricity trade with countries like Bhutan, Bangladesh and Nepal, also committed to hydropower projects. India boosts the economic integration among the neighbouring countries as with the South Asia Free Trade Agreement (SAFTA) for trade liberalization with the neighbouring countries, supports entrepreneurship and employment in the region. Improving connectivity and undertaking tariff reduction and other trade facilitation measures for expanding trade with SAARC offers another major area with considerable potential for mutual gain for India given that intra-regional trade is much below its potential in the region and expanding of markets provides India and all other South Asian countries with large untapped benefits (Chaturvedi, 2021: 43). The engagement with South Asia regions under India's initiatives foster the digital infrastructure development assistance to neighbouring countries, encouraging of participation as well as promoting investments in infrastructure, technical and strategic alignments of expansion sectors like media platforms, broadband in neighbouring countries. The South Asia Satellite represents an illuminating case of how India's use of space technology is increasingly becoming intertwined with political and international imperatives, reflecting the broader dynamics in the region (Stroikos, 2024). India's regional policy was underpinned by the goal of preserving its pre-eminence in South Asia and ensuring that the Indian sub-continent remained its exclusive sphere of influence by limiting extra-regional intervention (Destradi, 2012).

(*d*) *Conservation of Environmental Resources* (*SDGs 13-15*): India contributes its potential in South Asia regions to foster SDGs in better and significant status with the presence of climate security, sustainability and responsible governance. India meets with the potential of optimistic opportunities as the nation subscribes to the initiatives of the Paris Peace Agreement, ensuring conservation of the environment. It enhances the green energy and forest protection programmes with neighbouring countries like Bhutan, Bangladesh and Nepal, and arranges disaster management initiatives to mitigate any environmental challenges. India plays a critical leadership role for other emerging markets and developing economies (EMDEs) in the Global South and will demonstrate this through its upcoming G20 Presidency in 2023 and by having co-founded initiatives like the International Solar Alliance, One Sun One World One Grid and the Coalition for Disaster Resilient Infrastructure (Choudhary, 2022). It even promotes SAARC blue economy with the member countries and control pollution in South Asia regions, exploring exclusive economic zones and offer mutual benefits from it. The campaign of maritime security with Sri Lanka and Maldives recognised as an important role in terms of SDGs achievement in the region. Along with such initiatives, India engages for ecological conservation promoting afforestation projects with other South Asian countries, these include, for example forests (such as the Sundarbans requiring cooperation between India and Bangladesh), and the Thar ecosystem (requiring cooperation with Pakistan to halt desertification and conserve the fauna), (Chaturvedi, 2021: 45).

(*e*) *Strengthening Regional Partnership for Goals* (*SDG 17*): India's leadership to achieve SDGs and promote regional cooperation ensures with the indicators of SDGs as the presence of multiple regional frameworks to which the state has ensured its activities and goals in a significant upgradation. The engagements with multiple arenas indulge the growth in technology, economic integration, medical assistance, humanitarian programme, infrastructure and investments related initiatives mitigating the challenges of cooperation in the region. As cooperation has been set up with the SAARC Development Goals adopted at the Thirteenth SAARC Summit based on the recommendations by the Independent South Asian Commission on Poverty Alleviation (ISACPA) in 2007. The report had envisaged 22 Goals with 75 indicators (Chaturvedi, 2021: 46). BIMSTEC is well positioned to make sizable contributions towards advancing regional connectivity, though a series of organizational and substantive reforms would help increase its effectiveness (Xavier, 2018). It comprises cohesive engagement in the multilateral forums, building infrastructure to enhance growth and providing a line of credit to fast track development in the region (Shylan, 2024).

India's Engagement with Regional Sustainability Initiatives in South Asia

India's Bilateral Initiatives for Regional Cooperation and SDGs

India's reputation and standing have experienced considerable changes. India has been the leading power in South Asia, actively engaged in bilateral and regional strategic initiatives to promote sustainability aligned to UN's Sustainable Development Goals (SDGs). The regional development and cooperation to India's foreign policy is guided by the ethical considerations of "Neighbouring First" and "Act East" Policy, which emphasize the connectivity, regional stability and mutual growth to a significant extent. The nation's approach to bilateral cooperation linked with SDGs in several elements, the initiatives of engagement with neighbouring countries are discussed below:

(*a*) *India and Afghanistan*

India has engaged its significant role with Afghanistan focusing on regional cooperation and attaining sustainability between the two nations, the interaction emphasized Afghanistan's development, particularly through infrastructure projects, heavy investment on trade and commerce and humanitarian programmes. India's approach to participation in Afghanistan has shown to be highly effective, garnering the respect and support of India and is seen by the Afghan government and people as a trustworthy economic partner and ally (Nair, 2015). India aligned its initiatives of SDGs to achieve in regional cooperation, emerges as regional donor to which invested significantly in various major developments, as India's total commitment for the reconstruction of Afghanistan is US$2 billion, with more than half disbursed. It has funded/co-funded three very important infrastructure projects, the construction of the 218 km road from Delaram to Zaranj which gives Afghanistan access to an alternate port, Chabahar in Iran; it has reconstructed and expanded the Salma Dam which would produce 42 MW of power and irrigate 75,000 hectares of land when fully commissioned by mid-2016; it has co-funded and built transmission towers over the Hindu Kush as part of the Northern Electric Power System (NEPS) that has brought electricity to Kabul and other areas (Sinha, 2017: 138). The country has focused on pharmaceuticals, clothes, grains, dairy products to export which maximize the potential of their engagement in a manner of sustainability along with development in education and research initiatives ensuring non-governmental organisation to take part and promote the goals of SDGs. The instability in the region has, however, affected the relations but also flourish by time, to meet the achievements of SDGs, India has arranged; to attend to the massive and urgent medical needs, rushed a team of 13 doctors and paramedics to Kabul in the end of 2000, installing camps for fitting artificial limbs were held in different parts of Afghanistan throughout 2002 and since then, five Indian Medical Missions (IMMs) have been working in Kabul, Herat, Jalalabad, Kandahar and Mazar-e-Sharif, attending and disbursing medicines to 30,000 patients per month; these five IMMs cater to the poorest of the poor patients, many of whom come for consultation and free medicines from the contiguous provinces (India and Afghanistan Development Partnership, 2016). Both the nations, have emerged as inclusive development in relation with regional stability and India seeks to enhance its standing as a prominent regional actor by actively participating in the Afghan economy and offering support to the government and populace (Khan, 2024).

(b) India and Bangladesh

The two nations share a long history of cooperation based on historical, socio-economic, cultural and several significant factors. Both engaged their partnership in multiple areas, which contribute significantly to achieving the SDGs in both nations and focusing regional cooperation into the perspectives of India. The engagement includes, as Bangladesh as India's largest trading partner in South Asia with duty-free access under SAFTA agreement which even align since independence as of mutual-aid, cooperation, extended non-repayable money etc. to Bangladesh for their economic reconstruction. In 2010, the two countries encapsulating their mutually shared vision for the future, which would include cooperation in water resources, power transportation and connectivity, tourism and education (Joint Statement by India and Bangladesh on first meeting of the India-Bangladesh Joint Consultative Commission, 2012), this often ensures the productivity of SDGs in the bilateral relations among the two nations. The Sustainability is essentially focus with the economic integration as emphasized with inclusive economic growth, industry and innovative infrastructure, investments with growth of trade and commerce and reducing disparities in accessing resources. The connectivity beyond borders deals with road, railway, inland waterways and air connectivity promoting regional integration. They have operationalized the agreement for the usage of Chittagong and Mongla ports in 2023, avail services for transit cargo between North-east and mainland India (Bilateral brief of India-Bangladesh, 2024). The successful operation of cross-border interconnectedness between India and

Bangladesh forms the groundwork for greater energy cooperation in the South Asian regions (Rehman *et al.*, 2021). As India have engaged with several programmes in Bangladesh exploring the strength of health sector, water resources management, border security, renewable energy, environmental conservation and many significant spheres of cooperation even though the presence of challenges still exists, but maintain to enhance the engagement with more cooperative and dynamics of strategic steps.

(*c*) *India and Bhutan*

One of the happiest nations, Bhutan plays a vital role in regional cooperation in South Asia, ensuring India's engagement with SDGs as the two nations holds a special relationship based on mutual respect and cultural affinity. As a landlocked country, Bhutan relies on India mostly, for its engagement, covering more than half of the contribution and FDI from India itself, India has often engaged with multiple arenas engaging with sustainability as it promotes its presence in developments like economic assistance, humanitarian programme, infrastructure and investments, trade and commerce, security collaboration and socio-cultural exchange. These ensures the productivity of sustainability to significant extent to regional prosperity and stability. Indian contribution to Bhutan's Five Year Plans mainly covers the Project Tied Assistance (PTA) which funds 71 mutually agreed projects like agriculture, ICT, education, medical sciences, media, health, roads, urban development, judiciary, civil aviation, energy, road safety and transportation, constitutional offices and media, labour and human resource development and so on (Choden, 2004). India also grants and supports hydropower generation and other infrastructures. These include the 1020 MW Tala Hydroelectric Project, 60 MW Kurichhu Hydroelectric Project and 336 MW Chhukha

Hydroelectric Project. In addition, Paro airport, Penden cement plant, major highways, electricity transmission and distribution system, Bhutan Broadcasting Station, Indo-Bhutan microwave link, survey and mapping and exploration of mineral resources are also funded by the Government of India (Lama, 2023). India has engaged with sustainability in Bhutan focusing on digital and communication configurations with fibre-optic networks for faster internet connectivity and cross-border digital payment systems, it has ensured presence in education, training and skill development and optimization to flourish the spectrum of tourism and cultural affinity for development, which ensures the indicators of SDGs significantly. An important part of this approach has been the focus on enhancing capability and developing infrastructure with the goal of strengthening Bhutan's economy and fostering sustainable development (Ministry of External Affairs, Government of India, 2020; Ministry of External Affairs, Government of India, 2022). The collaborative efforts between India and Bhutan demonstrate the importance of regional partnerships in addressing global environmental challenges; by leveraging each other's strengths and experiences, both nations aim to achieve sustainable development and environmental conservation goals and such marks a significant step forward in this enduring partnership, promising a greener and more sustainable future for both countries (India News Network, 2024).

(d) India and Maldives

The achievement of regional cooperation while ensuring the SDGs in the region, India promotes its engagement with Maldives in significant arrangement with close and multidimensional interests. India has engaged for major development and infrastructure projects, economic integration, sustainability and security cooperation and humanitarian assistance in the region, ensuring the goals of SDGs. As part of neighbouring first policy of India's initiatives; exports to Maldives include agriculture and poultry produce, sugar, fruits, vegetables, spices, rice, wheat flour, textiles, drugs and medicines, a variety of engineering and industrial products exploring the sustainability goals and enhancing the bilateral trade stands at US$ 223 million in 2018-19 as against US$ 152 million in 2014-15 (India-Maldives Economic Relation, 2019: 9). The regional cooperation is highly appreciated as of the inauguration of India funded water sanitation projects in Maldives, and ensures with the endeavours in reaching SDGs (NDTV, 2024). India has been the first responder during COVID-19 pandemic and even participate in several humanitarian assistance and programmes in stability in the region. The sustainability components offer with several capacity building and training development with SAARC fellowship, technical cooperation and medical assistance in education and research, restoration of tourism and connectivity in support and mostly significantly the relations boost with presence of India-Maldives Free Trade Agreement, operational since 2011.

(*e*) *India and Nepal*

They share a deep-rooted tie, reinforces by the treaty of peace and friendship in the year 1950. Their bilateral cooperation lens with multiple arenas engaging with trade and commerce, energy efficiency, health and infrastructure, education and disaster management, which flourish India's engagement to SDGs achievement in South Asian region. This cooperation aligns as India been largest foreign direct investment in Nepal, introduces trade agreements and infrastructural initiatives for cross-border connectivity and employment generation, their presence of Integrated Check Posts which facilitate trade and commerce among the two nations, support to finance in services like banks, skill development, opportunities and inclusive growth in the region has been articulated by India since long which even boosted by humanitarian and sustainability projects. In addition to the relationship between Delhi and Kathmandu, Nepal's economic, cultural and social links with the Indian States of Uttarakhand, Uttar Pradesh, Bihar, West Bengal, Sikkim, and Bihar share a long border with Nepal while Raxaul and Jogbani are among the most important transit routes between India and Nepal, these Indian states not only share similar ecosystems and agro-climatic conditions with Nepal (Saran, 2023) which are considered as essential in terms of sustainability advancement within the two countries and also to mention; the first broad gauge passenger rail service between Bihar and Nepal started in 2022, both the states started the first cross-border oil pipeline in South Asia with a 69-km-long petroleum pipeline between Motihari in Bihar and Amlekhgunj in Nepal's Bara district in 2019 (Saran, 2023) ensuring a new step in their friendly relationship. The engagement of India and Nepal ensures with establishing renewable energy projects for reducing dependency in fossil fuels and electricity connectiveness in the surface of SAARC and BIMSTEC flag, India has been approachable towards the United Nation's SDGs in a significant exposure. India is planning to expand the BBIN electricity grid to include Southeast Asian countries that could emerge as a unified market, for Nepal, BBIN could also facilitate multi-modal transport opportunities *via* road, rail and inland waterways and would not only reduce costs but also the carbon emissions (Haran, 2018). The idea of BIMSTEC is woven around the 'Bay of Bengal' as a common neighbourhood, which could help landlocked countries like Nepal to further, increase their maritime trade and find connectivity to Southeast Asia. The increased and diversified economic connectivity would increase exports and have the potential to strengthen tourism sector of both the countries by attracting Buddhist pilgrims from Myanmar, Sri Lanka and Southeast Asian countries (Mehndiratta & Nora, 2022).

(*f*) *India and Pakistan*

Since the emergence of India and Pakistan as independent states, both the nations have experienced complex and intense relationships. Although, they have similarities in certain dimensions, Pakistan has been the only nation in South Asia, to which India has engaged only a minimum interest in bilateral relationships even though the presence of strong position of friendship holds with engagements of several humanitarian programmes, water resources management, environmental conservation, global platform engagement and health sector but the relationship tastes bitter because of past four wars and rivalry which even destabilizes the regional peace and security. The bilateral cooperation with Pakistan can be articulate with the trade and economic prosperity as India exported goods of $2 billion to Pakistan, which was only 0.6 per cent of all Indian exports and in terms of imports, India took in $495 million worth of goods from Pakistan, 0.096 per cent of total imports back in fiscal year of 2018-19 (Sami, 2024) before declining the economic cooperation between the two nations, the trade route at Wagha-Attari remains a key corridor for commercial exchanges; ensures the productivity of sustainability in a sense. The facilitation of SAARC and South Asian Free Trade Area (SAFTA) is not much possible towards the engagement of the two nations, but India arranged to provide its cooperative arrangements to almost all other nations in the region. The Indus Water Treaty (1960) governs the sharing of the Indus River and its tributaries, the nations arranged to look after the commission for hydropower projects. India urged to build a corridor with suitable facilities in its territory from the international border to Gurdwara Kartarpur Sahib in Pakistan to facilitate easier access and smooth passage of Indian pilgrims (India-Pakistan Bilateral Relation, 2020) dealing with the engagement of the two nations. Moreover, the successful completion of TAPI is expected to enhance regional energy connectivity, delivering economic benefits to all participating countries while meeting growing energy demands across South Asia (Mohammadi, 2025) which also brings India and Pakistan on a common platform of cooperation. India cooperated in SAARC COVID-19 emergency fund during the pandemic and provided essential medicines and humanitarian facilities to Pakistan ensures the indicators of SDGs, it also provides scholarships under ICCR (Indian Council for Cultural Relations) programmes to the Pakistani students while promoting people to people exchange, engaged humanitarian efforts during Russia-Ukraine War in 2022 and even negotiates for climate challenges and conservation management initiatives.

(g) India and Sri Lanka

Another country for regional cooperation in South Asia, keeping India's perspective in mind would be Sri Lanka as both the nations have a close relationship which emerged with engagement in multiple sectors, as an immediate neighbour and deep linkages in several elements, the two nations experience an immense cooperation in the South Asia region. To promote the regional development and optimization of successful SDGs, India enhances its active participation in Sri Lanka relying within several indicators which evolved into multifaceted dimensions covering sustainable economic integration and trade (Business Standard, 2024), infrastructural initiatives and investments, health and ecological upgradation and stabilizing the development of Sri Lanka with technical, medical and humanitarian assistance. These included a USD 400 million-currency swap *via* the Reserve Bank of India's (RBI) South

Asian Association for Regional Cooperation (SAARC) swap facility and the deferral of about USD 500 million in trade liabilities accumulated through the Asian Clearing Union (ACU), primarily from imports from India (Panduwawala, 2024: 89). FDI inflows from India have shown less volatility, with a greater number of individual investments due to the substantial presence of Indian firms operating across various sectors in Sri Lanka (Panduwawala, 2024: 99). As being the partner of SAARC, both the countries engaged with multiple arenas for cooperation and strategic dynamics in the region, with the Free Trade Agreement since 2000 the trade and commerce emerged with high potential, working jointly on several projects like rehabilitation of Kankesanthurai port, Digital identity, regional energy and industrial hub, Jaffna-Colombo rail connectivity and many other allies of engagement. The sustainability engagement has also articulated with energy cooperation with Sri Lanka as well as maritime security and defence assistance ensuring SDGs to achieve significant growth and resemblance with development for future opportunities of both the nations.

India's Engagement with Paris Peace Agreement and Implications on South Asia

The role of an expanding equity-based discourse coalition in maintaining India's preferred framing of climate change was particularly evident at the first Conference of the Parties (COP-1) in 1995 (Stevenson, 2011). India plays an important role in multilateral climate negotiations to which it comes with a vision of participating in a positive and dominant role in climate change negotiations. The Paris Agreement on Climate Change shows the strong commitment of the international community to address the issue of Climate Change. The seed of the Paris Agreement (hereafter referred to as the Agreement) was sown at the 17th Conference of Parties (COP) Decisions held in Durban in 2011. The Paris Agreement aims to limit global temperature rise to well below 2°C and to make best efforts to hold warming to 1.5°C above pre-industrial levels (UNFCCC, 2020). The Agreement is guided by the principle of equity and common but differentiated responsibilities and respective capabilities (CBDR-RC) in its pursuit to attain the objectives enshrined in the UNFCCC. India's ratification of the Agreement is a significant step of reaffirmation to address the climate change, which even boosted its relationship with neighbouring countries, focusing growth to least developing countries and mutual benefits in the big picture. India ratified the Paris Agreement in 2015 showcasing itself as a responsible global player. In so doing, it bore in mind its development agenda, particularly poverty eradication and provision of basic needs for its citizens, coupled with its commitment to follow the low carbon path to progress, and it was guided by the assumption of unfettered availability of cleaner sources of energy and technologies and financial resources from around the world, apart from having trust in a fair and ambitious assessment of global commitment to combating climate change; the Indian government has declared its support for the Paris Agreement through its readiness to become a signatory for a sustainable, low-carbon future. India has laid a strong groundwork to stand up for greater global cooperation on climate action through its pledges (Ratha, 2020). India has been positioned with a strong determination by pressing the segment of "climate justice" to which the importance of developed countries is determined to reduce the greenhouse gas emissions and provide support to developing countries to embrace renewable energy. India's position on climate change negotiations will have far-reaching implications for the success of global climate cooperation (Gupta *et al.*, 2015). The engagements with Paris Peace Agreement ensure with exploring energy solutions, cross-border connectivity, determining regional frameworks with BIMSTEC energy grid, hydropower projects with neighbouring countries and SAARC Disaster Management initiatives to strengthen resilience in the region. India has also committed this with national sustainable development with proper mechanisms, ensuring its motive of achieving SDGs by exploring the potential in South Asia regions for climate-resilient and sustainable future. As India has been an engaged stakeholder to the Paris Agreement, and this commitment helps confirm that India is dedicated to increasing ambition on climate change (The Road from Paris, 2020).

India and United Nations Development Programme (UNDP) Initiatives in Regional Cooperation

India being a rising power in international community, also a founding member of United Nations, has emphasized its partnership with United Nations Development Programme (UNDP) for ensuring sustainable development policies, basically both in national and international levels. These ensures to engagement of collaboration extensively to mitigate the challenges related to economic integration, climate resilience, gender equality, poverty alleviation, infrastructural initiatives and humanitarian assistance. The India-UN Development Partnership Fund is another practical manifestation of climate action and sustainable development in the spirit of South-South Cooperation. The Fund established in 2017, aims to contribute to the efforts of developing countries towards the realization of the 2030 Agenda for Sustainable Development Goals including on Climate Action (India and United Nations, 2022). The country's engagement with UNDP has implications with SDGs goals and promotion of regional cooperation in South Asia regions, as the location suffers vulnerable challenges and SDGs-led-initiatives would emerge to mitigate those, and preserve a better future; align with developmental priorities and global mandate. The India-UN Fund is a notable example of South-South Cooperation with the multilateral system, sponsored by the Government of India and implemented through the United Nations system, with the United Nations Office for South-South Cooperation (UNOSSC) as the coordinating partner, the Fund supports projects that advance all Sustainable Development Goals (SDGs); the priority is accorded to partnering with small island developing States, least developed countries, landlocked developing countries and countries affected by disaster (India-UN Development Partnership Fund, 2021). With the support of UNDP, India ensures its growth in multiple sectors, which serves as models for South Asia nations in collaboration with SAARC and BIMSTEC initiatives, for sustainability and responsible regional stability. The engagement highlights the presence of joint renewable energy projects, climate adaptation and disaster management in the region, sharing of knowledge and exchange of people in sustainable development, India cooperates with cross-border women entrepreneurship in the region for growth of gender equality especially advancing SAARC frameworks, to ensure the SDGs, the South Asian regions experienced positive response from India as being largest provider of its stronghold in several elements, the region captivating its strength of development policies through several bilateral and multilateral forums of engagement. India's plans to strengthen the mechanisms for SDGs monitoring and reporting through a dashboard currently under development, with technical support from the UNDP (Voluntary National Review Report, 2017) has ensured the growth in South Asian regions for the cooperative growth and resemblance with development of sustainable strategy.

International Solar Alliance (ISA)

The International Solar Alliance (ISA) is one of the India's most ambitious global initiatives, which has potential for adaptation in renewable energy and sustainable development. The International Solar Alliance (ISA) is a global initiative launched in 2015 by India and France at the COP21 summit in Paris to promote solar energy as a sustainable solution for energy access and climate change, headquartered in India, the ISA is the first international organization established in the country, reflecting India's commitment to multilateralism and a carbon-neutral future (Ministry of New & Renewable Energy, 2024). The South Asian regions, where the focus for cooperation in certain dimensions is required to engage the countries in more effective way, India comes forward with a lamp of opportunities to accelerate SDGs especially in climate, industry, innovation and infrastructure. ISA helps countries with insufficient access to energy by giving them a voice on the international platform to coordinate collaborative efforts and share ideas in order to close technological gaps and create partnerships across the globe (Mishra, 2024). The ISA launched several key initiatives to advance global solar energy adoption. The SolarX Start-up Challenge, introduced at COP27, supported innovative solar businesses in ISA Member Countries. The STAR-C Initiative strengthened solar technology skills in developing economies, the Viability Gap Funding Scheme provided grants to solar projects in Least Developed Countries and Small Island Developing States, easing financial barriers (Ministry of New & Renewable Energy, 2024). India has developed the "One Sun, One World, One Grid" (OSOWOG), which will be a transnational electricity grid supplying solar power across the globe (PIB, 2018) which ensures to enhance connectivity with the first phase will connect the Indian grid with the Middle East, South Asia, and South-East Asia grids to share solar and other renewable energy resources;

the second phase will connect the countries in the first phase with the African pool of renewable sources; and finally, the third, concluding, phase will be one of global interconnection (Jai, 2020). The ISA has engaged with India as initiatives for solar microgrids and decentralised solar alliance solutions to bridge the gap among South Asian regions, ensuring clean and affordable electricity, technology transfer, research and larger solar production in the region; cheaper solar installations with India's expertise. These often considered as impactful for employment generation and bigger prospects of opportunities to the region, improving regional stability and mutual development. India has maintained its long-standing position on the "differentiated responsibility" of developing and developed countries, and particularly that climate action in the developing world hinges on adequate funding and technology transfer (Jha, 2023).There is great optimism that the solar alliance would help in pushing 'power for all' into a more realistic realm (Bhattacharya *et al.*, 2018). India's leadership in the International Solar Alliance plays dynamic role in SDGs achievement in South Asian regions, with a stronger regional commitment and paradox of sustainability.

Coalition for Disaster Resilient Infrastructure (CDRI)

India-led climate adaptation initiative with Coalition for Disaster Resilient Infrastructure (CDRI) that showcases India's rising leadership role in making climate action and disaster resilience an integrated and inclusive approach. Infrastructure systems such as highways, power stations, communication lines, and even housing are the key drivers of sustainable development, economic growth, and prosperity; these modern infrastructure projects are regularly exposed to extreme weather events such as turbulent rains, massive floods, and historic heat waves which call for collective response and action. It enhances to promote resilient infrastructure systems with climate change impacts. The region of South Asia, is one of the prone areas for disaster in the world, the leadership of India in CDRI holds a significant step for achieving the SDGs. The Coalition for Disaster Resilient Infrastructure (CDRI) aims to enhance global understanding and action on climate and disaster resilient infrastructure (DRI) through knowledge creation, curation and dissemination (Sablok, 2023).The formation of the Coalition was a result of a series of national, regional, and global consultations with countries that faced a variety of challenges posed by development, climate, and disaster risk factors. India provides its technical expertise and financial assistance to help its neighbours build resilient infrastructure, ensuring long-term stability and sustainable development. Generally, India launched it in 2019, to address the disaster resilience which offering more than forty countries, globally. The Infrastructure Resilience Accelerator Fund (IRAF) supported by the country, helps South Asian regions towards the construction of connectivity and urban development. The engagement promotes capabilities of India as like financial and humanitarian assistance during crisis in neighbouring countries like Nepal (earthquake, pandemic), Bangladesh and Myanmar (cyclone, floods, political instability) and so on. The regional framework of BIMSTEC Disaster Management Centre indulge with India's vision and tactical leadership to reach SDGs initiatives in the region, the presence of National Disaster Response Force (NDRF) explore the potential of India's strong determination in the region, ensuring SDGs in South Asia.

Global South Advocacy

South Asia stands with challenging regions for SDGs achievement, the presence of India's leadership role in G-20 has direct impact for regional cooperation and transformative growth in the region. India hosted the G-20 Leaders' Summit on 9-10 September, 2023, bringing together the G-20 Heads of States and Governments on the theme of "One Earth, One Family, One Future, *Vasudhaiva Kutumbakam*" (Chaturvedi & Kumar, 2022), which focused on economic development, digital configuration and climate change with the implications of SDGs in South Asia. India has always been recognised as the Voice of the Global South, which includes African nations also (Kumar, 2024). India maintains a stance that intentionally does not clarify which countries are included in the scope of the Global South and keeps the definition vague (Moriguchi, 2023). India's initiative for SDGs achievement relies with Global South to which the debt relief advocacy was heavily introduced along with trade liberalization, digital integration to help South Asia region as more economically resilient area. This ambiguity indicates the possibility of including all countries other than developed countries and China, and as a result, India seems to have succeeded in bringing into view the Global South, a group of emerging and developing countries, as a single political actor (Bolt, 2023). The country has taken significant steps to mitigate greenhouse gas emissions, promote renewable energy, and strengthen resilience to climate impacts; India actively participates in global climate negotiations and initiatives, advocating for the concerns and interests of developing nations in the Global South (Rekhi, 2023).

Challenges in Promoting India's Role for SDGs in South Asia

Although, India's engagement with SDGs in South Asian regions have articulate as emerging development and even designated with success, to stabilize the cooperation in the region. However, there are several challenges which are determined to be mitigate with changing times, as the problems have a huge impact in the region's development and optimization of implementation to SDGs. Some of the key challenges are discussed below:

(*a*) *Political and Diplomatic Tensions*

There is a long-standing conflict in South Asia particularly between India and Pakistan on geographical factors, these hampers the strategic growth in multiple dimensions in South Asia, both the countries are engaged in SAARC, also the presence of cross-border terrorism and political disagreements has led to backlash the opportunities in the region. The clashes among the countries with emerging global power dynamics, and rising of China's initiative in the region has also somewhere reflected negative impact, for India's role of engagement to neighbouring countries for sustainability and mutual development.

(*b*) *Bilateral Disputes*

The region suffers from bilateral disputes, as covering a vast land surface, the countries often get themselves engage in controversy for their disagreements in certain initiatives. For instances, in case of India, the country has experienced bilateral disputes with Bangladesh for river sharing, regional security and flow of migration, Nepal for border sharing and Sri Lanka relying with water sharing and fishing aligns. This often impacts the India's vision of SDGs for regional cooperation in South Asia.

(*c*) *Structural Nature of Organisations and Its Functioning*

The regional organizations and frameworks presence in South Asia regions, like SAARC has often considered its failure in terms of engagement in cooperative manner for offering socio-economic integration due to political rivalries among India and Pakistan, the implementation of economic cooperation with SDGs initiated by SAFTA has also not positioned in correct order, also BIMSTEC has not engaged in active status to mitigate the challenges and ensure integration with the South

Asian countries, slow progress of BIMSTEC and BBIN on connectivity is affecting regional transport infrastructure and limited enhancement has determined a big challenge for South Asia regions growth and resemblance to India's engagement.

(*d*) *Economic Disparities and Integration*

The presence of unequal economic integration among the countries of South Asia, has also affected the SDGs goals in the region. The rate of exports and imports are often lower as compared to Integration of Trade and Commerce to other parts of the world, even after Free Trade Agreements, the region has not introduced itself into high potential cooperation, rises unemployment, minimum industrialization and lack of modern infrastructure. India has introduced several projects to meet the requirements of SDGs, initially slow in progress and even such developments, the poverty remains a significant barrier among the countries of the region.

(*e*) *Environmental Degradation*

There is presence of need of environment preservation in the region, although several programmes and initiatives has engaged by India in South Asia region but, the region needs more enhancement to indulge productive growth with vulnerable climate change in the region. The cooperation on disaster management, rising sea levels, frequent climate change affecting trade and commerce and even with disputes of water sharing among the countries like Teesta River conflicts and Indus Water conflicts lead to slow progress in achieving SDGs in the region. The renewable energy projects are initial in the regions, but face financial delays and administrative ignorance to implement it effectively with neighbouring countries of India.

(*f*) *Limited Connectivity and Rising Migration Issues*

The South Asian regions explore with great leadership of India when it comes to investments and infrastructure, but often face challenges in terms of weak transport infrastructure affecting SDGs in innovative and industrial growth. The administrative delays in implementation of projects and slow diplomatic results ineffective growth in the region for development in higher oriented segments, the region also suffers from low digital infrastructure and hereby hampers to sustainable environment. Again, the migrations among the South Asian regions, creates challenge to achieve SDGs, face major inflation and refugee crisis such as Rohingya, Afghan crisis and recent crisis of Bangladesh in 2024. The cross-border migration between India and neighbouring countries have also raises issue of national security and human rights implications.

Recommendations

In order to strengthen India's role in promoting regional cooperation for achieving Sustainable Development Goals (SDGs) in South Asia, and ensuring the importance of SDGs for development and engagement in multiple dimensions, the following are some important recommendations that are mentioned keeping in view about India's engagement with sustainability and responsible development in the region. These follows:

First, India should engage in strengthening regional institutional mechanisms. As the South Asian countries have engaged with multiple arenas indulging its cooperation with neighbouring countries, India as an emerging economic power, should ensure its engagement towards regional institutional mechanisms focusing the objectives and developmental priorities of SAARC, BBIN and BIMSTEC particularly arranging dynamic possibilities of discussion with SDGs achievements and its effective mechanisms. With the development cooperation strategy, India should engage with regional forums to build trust and mutual growth, as mitigating the emerging challenges flourishing the founding principles of India's foreign policy. It is possible only with sustainable relationships with neighbouring countries that India would ensure its presence in the region and develop SDGs.

Secondly, the regional sustainable infrastructure development is essential for achieving SDGs in the region. The key indicators like clean energy, education, poverty eradication, renewable energy, infrastructural development initiatives and regional enhancement should be articulate and focus for the development. With the potential of optimistic opportunities, India should engage in profound capabilities for ensuring systematic engagement more strategically, interested grounds are needed to be summarised and this ensures the productivity of India's regional cooperation. The projects and initiatives of development like BBIN Motor Vehicles Agreement, India-Bhutan Hydropower Agreement, India-Bangladesh Electricity projects and so on, are required to focus more and complete them within a time-frame. These initiatives are essential segments, which are important for regional sustainability and mutual growth.

Thirdly, the region of South Asia, experiences several climate issues to which India can promote its leadership and expertise to mitigate such challenges and explore the opportunities of sustainability and regional security. India could ensure its presence with fund allocation dealing with disaster orientations, humanitarian assistance, and financial aids to affected countries. The expansion of the International Solar Alliance was a great opportunity for India, towards which the nation should ensure its presence more and make necessary arrangements into a clean energy transition. India should keep its commitment to renewable energy, low carbon economy; investments on energy efficiency along with should engage North-eastern part of India for suitable enhancement with economic integration and emphasis climate preservation focusing on regional cooperation.

Fourthly, India should focus on public health collaboration, even during the COVID-19 pandemic, the state has been capable of serving an effective role in implementation of vaccine diplomacy. The nation has vast pharmaceutical strength, to which it should arrange in productive networks to neighbouring countries by providing medical assistance and health orientations to meet the requirements. India should engage its role in SDGs in terms of both health and educational sectors, as of South Asian University, to facilitate collaborative research on sustainability and healthcare challenges in the region, which would ensure sustainable development.

Lastly, India should ensure its productivity with digital transformation and infrastructural initiatives to neighbouring countries. The presence of high-tech quality and economic prosperity in South Asia, India should arrange its potential beyond borders and engage digital mediums with members of SAARC for regional sustainability and mutual growth. It has expanded the India UN Development Fund to finance digital literacy initiatives and technology-based SDGs solutions, it should arrange to indulge the countries even more for benevolence growth. The state should engage infrastructural initiatives to promote regional cooperation in neighbouring countries, ensuring investments on multiple dimensions to ensure prosperity and stability in the region.

Conclusion

South Asia as a region is one of the important geographical locations in the world, which ensures the need of sustainable development. The region is engaged with multiple dimensions of regional cooperation and ensures the productivity in accelerating the healthy growth. There has been a structural transformation of the country's economies, since representing the colonial strata and then enhancing development in contemporary times, presenting a huge gap of sustainable development as compared to Western world. The region has been recognised with numerous challenges of poverty, unemployment, climate change, poor infrastructure and limited growth in the economy. However, India emerges as the largest influential country, has driven potential towards the development of sustainability by relating regional cooperation, economic integration, investments in infrastructural development, socio-economic progress, and climate resilience. The challenges of the region are focused on the perspectives of India, where it promotes capabilities to regional cooperation and achieve SDGs, which are essential priorities to achieve and maintain stability in the region for efficient and inclusive growth. The sustainability engagements are prime concern with India's vision, relating to its regional cooperation and strategic foreign policy initiatives that are mutually organised to achieve its targets and ensures a benevolence growth in the arena of cooperation. These initiatives are determined with regional stability and willingness to share frameworks of SAARC and BIMSTEC particularly to raise the productivity of sustainability, engaging several programmes of climate change, infrastructure development, energy efficiency, health and educational sectors. The global initiatives articulated with the Paris Peace Agreement and UNDP, India indulged its voice to replicate the challenges and explore opportunities for Global South and its affairs, the state has expanded its potential beyond borders by enhancing the maritime security, cross-border connectivity, International Solar Alliance and several parameters of cooperation relating to fund orientations, humanitarian assistance and climate resilience respectively. Hence, the commitment explored various dimensions of challenges in the region and its implications highlight the serious factors, which are yet to be explored. The presence of strong problems relating to bilateral, political, economic and infrastructural development, India with its vision of foreign policy has tried to amplify the growth and resemblance to sustainability. To conclude, India with its initiatives of sustainable development, having tremendous development in engagement with SDGs, its commitment for 2030 Agenda ensures sustainability for South Asian regions. The success of its cooperation deals with the determination and strategic steps that would ensure the growth of the country and its prospects in the region, keeping attributes to the potential of achieving SDGs for a better future.

REFERENCES

Asian News International, *Jaishankar Inaugurates India-funded Water, Sanitation Project in Maldives.* 11 August. https://www.ndtv.com/india-news/eam-jaishankar-inaugurates-india-funded-water-and-sanitation-project-in-maldives-6310871

Bhattacharya, Shreya, Ekta Niranjan, and Chithra Purushothaman (2018). "India and International Solar Alliance." (Institute for Defence Studies and Analysis).

Bolt, Neville (2023). "Why Strategic Ambiguity is so Ambiguous." (Defense Strategic Communications) 12. https://stratcomcoe.org/publications/defence-strategic-communications-volume-12-spring

Business Standard, *India to Continue Backing Sri Lanka's Sustainable Economic Growth: Murmu.* 16 December. https://www.business-standard.com/external-affairs-defence-security/news/india-to-continue-backing-sri-lanka-s-sustainable-economic-growth-murmu-124121601350_1.html

Chakradeo, Saneet (2020). *India Bhutan Hydropower Cooperation: Perceptions and Politics.* 11 November. https://csep.org/blog/india-bhutan-hydropower-cooperation-perceptions-and-politics/

Chattopadhyay, P. (2008). "Balancing Global Integration with Regional Cooperation: The South Asian Experience." *The Indian Journal of Political Science, LXIX*(3).

Chaturvedi, Sachin (2021). *Evolving Indian Strategy on SDGs and Scope for Regional Cooperation.* South and South-west Asia Office, New Delhi: Economic and Social Commission for asia and the pacific.

Chaturvedi, Sachin, and Kumar, Sushil (2022). "India's G20 Presidency: Pushes Global Economy Ensures Space for Global South." *Indian Foreign Affairs Journal, 17*: 146-157. http://www,jstor.org/stable/10.2307/48772506

Chaudhary, Anasua, Basu, Pratnashree and Bhonsale, Mihir (2015). *Driving Across the South Asian Borders: The Motor Vechicle Agreement Between Bhutan, Bangladesh, India and Nepal.* ORF PCCASIONAL PAPER#69, New Delhi: Observer Research Foundation.

Choden, T. (2004). "Indo-Bhutan Relations Recent Trends: Perspectives on Modern South Asia—A Reader in Culture, History and Representation." Cambridge University, pp. 298-302.

Choudhary, Kamya (2022). *How is India Tackling Climate Change?* 2 November. https://www.lse.ac.uk/granthaminstitute/explainers/how-is-india-tackling-climate-change/

Confederation of Indian Industry, *India-Maldives Economic Relations, A CII Paper.* New Delhi: Confederation of Indian Industry, 9.

Destradi, Sandra (2012). "Indian Foreign Policy and Security Policy in South Asia." *Regional Power Strategies*, Routledge.

Dhakal, S., Srivastava, L., Sharma, B. Nepal, R. and Wakhley, K.B. (2019). *Meeting Future Energy Needs in the Hindu Kush Himalaya.* Spinger, pp. 166-207.

Dutt, Sagarika (2023). "Peace and Development in South Asia: Problems and Prospects." *South Asian Survey* (Sage Journals) *30*(1): 123-141. doi:10.1177/09715231231204423

Ginwal, Richa, Kumar, Daleep and Verma, Sarika (2022). "BIMSTEC: Future Potential in Shaping India's Foreign Policy." *Zeichen Journal, 8*(11): 71.

Gupta, Biswanath, and Raju, K.D. (2016). "Space exploration by India and Socio-economic Cooperation with SAARC Countries." *India Quarterly*, Sage Publications, *72*(3): 278-289.

Gupta, Himangana, Kohli, Ravinder Kumar and Ahluwalia, Amrik (2015). "Mapping Consistency in India's Climate Change Position:

Dynamics and Dilemmas of Science Diplomacy." *Ambio, 44*: 592-599.

Haran, V.P. (2018). "Regional Cooperation in South Asia." *Indian Foreign Affairs Journal, 13*(3).

Hassan, M.K. (2001). "Is SAARC a Viable Economic Block? Evidence from Gravity Model." *Journal of Asian Economics, 12*(2): 263-290.

Hussain, Syed Faraz, and Zoya Abrar Khan (2023). "International Solar Alliance: The Geopolitical Imperative for India." *International Journal of Research Publication and Reviews, 4*(7): 2130. doi:https://doi.org/10.55248/gengpi.4.723.49443

India News Network, *India-Nepal Energy Partnership: Arun-3 Hydro Electric Project achieves significant milestone.* 6 May. https://www.indianewsnetwork.com/en/20240606/india-nepal-energy-partnership-arun-3-hydro-electric-project-achieves-significant-milestone?utm_source=chatgpt.com

Jai, S. (2020). *One Sun, One World, One Grid: All You Need to Know About Mega Solar Plan.* 15 August. https://business-standard.com/article/current-affairs/onesun-one-world-one-grid-all-you-need-to-know-about-solar-strategy-120081500417_1.html

Jha, V. (2023). "International Solar Alliance: Bridging the Gap." *Tracks to Transition: India's Global Climate Strategy*, Centre for Social and Economic Progress, pp. 37-43. httos://csep.org/Shselvt

Khan, Ovais Hussain (2024). "India and Afghanisthan: A Relationship that has been Evolving for Decades." *International Journal of Research in Humanities & Soc. Sciences, 12*(1).

Khatun, Fahmida (2019). *BIMSTEC Countries and Climate Change: Imperatives for Action.* ORF Issue brief No. 299, Observer Research Foundation, 21.

Khatun, Fahmida, and Syed Yusuf Saadat (2020). "Regional Cooperation for SDG Implementation: A Model for South Asia." Centre for Policy Dialogue, *3*.

Khurana, Gurpreet S. (2018). *BIMSTEC and Maritime Security: Issues, Imperatives and the Way Ahead.* National Maritime Foundation.

Kim, Misu (2024). "Great Transition in India's Regional Cooperation: Beyond the Neighbouring Countries." *The Journal of Indian and Asian Studies, 5*(2). doi:10.1142/S271754132400035

Kumar, Suneel (2020). "Reinvigoration of BIMSTEC and India's Economic, Strategic and Security Concerns." *Millennial Asia*, Sage Publications, *11*(2): 197. doi:10.1177/0976399620925441

Kumar, Sushil (2024). "India's G20 Presidency as a Voice of Global South." Research and Information System for Developing Countries.

Kumar, V. (2017). "Sub-regionalism in South Asia: A Case Study of the Bangladesh-Bhutan-India-Nepal-Motor Vehicle Agreement." *Strategic Analysis, 41*: 1-13.

Lama, Uttam (2023). "India and Bhutan: A Relationship before and After Independence." *ISEC Working Paper No.560* (Institute for Social and Economic Change).

Li, R., Wang, Q. Liu, Y. and Jiang, R. (2021). "Per-capita Carbon Emissions in 147 Countries: The Effect of Economic, Energy, Social and Trade Structural Changes." *Sustain Product Consumption,* pp. 1149-1164.

Mehndiratta, Shomik and Nora, Erik (2022). *What will it take to connect Bangladesh, Bhutan, India, Nepal (BBIN) Sub-region?* 24 January. https://blogs.worldbank.org/endpovertyinsouthasia/what-will-it-take-connect-bangladesh-bhutan-india-nepal-bbin-sub-region

Ministry of External Affairs, *India and Afghanistan: A Development Partnership.* New Delhi: Ministry of External Affairs, Government of India. http://meaindia.nic.in

Ministry of External Affairs, *India-Pakistan Bilateral Relations.* New Delhi: Ministry of External Affairs, Government of India.

Ministry of External Affairs, *India-Bhutan Relations.* Ministry of External Affairs, Government of India. https://www.mea.gov.in/Portal/ForeignRelation/Brief_Bilateral

Ministry of External Affairs, *Joint Statement by India and Bangladesh on First Meeting of the India-Bangladesh Joint Consultative Commission.* New Delhi: Ministry of External Affairs, Government of India. http://mea.gov.in

Ministry of External Affairs, *Keynote Address by Secretary (East) at the BIMSTEC Coastal Security Workshop.* Ministry of External Affairs, Government of India. https://www.mea.gov.in/Speeches-Statements.htm?dtl/32068/Keynote_address_by_Secretary_

Ministry of New & Renewable Energy, *A New Era in Solar Energy: Progress Driven by the International Solar Alliance.* New Delhi: Ministry of New & Renewable Energy, Government of India.

Ministry of New and Renewable Energy, PIB, *PIB, PM Inaugurates First Assembly of the International Solar Alliance.* New Delhi: Press Information Bureau.https://pib.gov.in/Pressreleaseshareaspx?PRI

D=1548295#:~:text=The%20first%20Assembly%2C%20in%20a,energy%20access%20at%20affordable%20rates

Ministry of Statistics and Programme Implementation, *Sustainable Development Goals, National Indicator Framework, Progress Report, 2024.* National Statistical Office, New Delhi: Ministry of Statistics and Programme Implementation, Government of India.

Mishra, Shivansh (2024). "Empowering the Global South: India's Role in Renewable Energy *via* International Solar Alliance." *International Journal of Research Publications and Reviews*, pp. 4604-4608. doi:https://doi.org/10.55248/gengpi.5.0724.2006

Mohammadi, Habib (2025). *Turkmenistan's President Orders Acceleration of TAPI Gas Pipeline Construction.* 8 February. https://amu.tv/156375/

Moriguchi, Jun (2023). *India's Global South Strategy-focusing on Diplomatic Relations with Africa.* Mitsui and Co. Global Strategic Studies Institute Monthly Report.

Nair, K.K. (2015). "India's Role in Afghanistan Post 2014 Strategy, Policy and Implementation." By C.D.Verma, 55. Centre for Land Warefare Studies.

Panduwawala, T. (2024). "India's Extraordinary Support during Sri Lanka's Crisis: Motivations and Impacts." *Journal of Indo-Pacific Affairs*, Air University, pp. 85-104.

Permanent Mission of India to the UN, *Permanent Mission of India to the United Nations: India and United Nations.* New York: Permanent Mission of India to the UN. https://pminewyork.gov.in

Prasain, Sangam (2024). *GMR Ropes in two Indian Partners for 900MW Upper Karnali Hydel Projects.* 12 February. https://kathmandupost.com/money/2024/09/13/gmr-ropes-in-two-indian-partners-for-900mw-upper-karnali-hydel-project?utm_source=chatgpt.com

Ratha, Keshab Chandra (2020). "India at the Paris Summit: A Potential Game Changer." *Indian Journal of Asian Affairs, 33*: 112-119. http://www.jstor.org/stable/10.2307/27003439

Reddy, Ramana (2020). "Indi's Role in the SAARC." *Journal of Emerging Technologies and Innovative Research, 7*(11).

Rehman, Saif, Shukla, Rahul, Alam, Kabir, Gautam, Aman, Baba, K.V.S., Narasimhan, S.R., Barpanda, S.S. and Soonee, Sushil K. (2021). "Success Story of India-Bangladesh Power Systems Operation." *9th IEEE International Conference on Power Systems* (*ICPS*). New Delhi: Power Systems Operation Cooperation Ltd. doi:10.1109/ICPS52420.2021.9670352

Rekhi, Vedika (2023). *India's Role in the Governance of the Global South.* 14 August. https://moderndiplomacy.eu/2023/08/14/indias-role-in-the-governance-of-the-global-south/

Sablok, Avni (2023). *Examining the Role of the Coalition for Disaster Resilient Infrastructure (CDRI).* 14 September. https://www.icwa.in/show_content.php?lang=1&level=3&ls_id=9924&lid=6344

Sami, Waheed (2024). *What are the Prospects for Reviving India-Pakistan Trade?* 17 May. https://thediplomat.com/2024/05/what-are-the-prospects-for-reviving-india-pakistan-trade/

Saran, Pankaj (2023). "India-Nepal Relations: Regional, Sub-regional and Bilateral Opportunities." *Centre for Research on Strategic and Security Issues* (NatStrat).

Sarkar, Sadeka, S., Feardous, A.A. and Ahmed, S. (2013). "Human Development Performance of Bangladesh." *Journal of Applied Science Research, South Asian Perspective, 9*(4): 2475-2484.

Scott, Christopher, Saklani, Udisha, Shrestha, Padmendra and Mukherji, Aditi (2020). "Hydro-energy Cooperation in South Asia: Prospects for Transboundary Energy and Water Security." Edited by Elsevier. *Environmental Science and Policy,* pp. 22-34. https://doi.org/10.1016/j.envsci.2020.07.013

Shylan, Prajwal (2024). *Engage South Asia—India's New Strategies for Regional Partnership.* 28 December. https://www.thegeostrata.com/post/engage-south-asia-india-s-new-strategies-for-regional-partnership

Singh, Amita (2022). "India's Role and Contribution to Building Global Resilience in Disaster Management." *Indian Foreign Affairs Journal,* Prints Publications Pvt. Ltd., *17*: 95-96. https://www.jstor.org/stable/10.2307/48745127

Sinha, Shakti (2017). "Rising Powers and Peacebuilding: India's Role in Afghanistan." In *Rising Powers and Peacebuilding,* by Charles Call and Cedric de Coning, 138. Cham, Switzerland: Palgrave Macmillan. doi:DOI 10.1007/978-3-319-60621-7

Sinha, Uttam (2024). *BBIN Sub-region: Perspectives on Climate-Water-Energy Nexus.* New Delhi: Pentagon Press, LLP.

Stevenson, Hayley (2011). "India and International Norms of Climate Governance: A Constructivist Analysis of Normative Congruence Building." *Review of International Studies, 37*(3).

Stroikos, Dimitrios (2024). "Space Diplomacy? India's new regional policy under Modi and the 'South Asia Satellite'." Routledge: *India Review*, pp. 46-70. doi:https://doi.org/10.1080/14736489.2023.2295715

Taneja, Nisha, Nag, Biswajit, Joshi, S., Rastogi, R. and Dua, S. (2024). "Bangladesh, Bhutan, India, Nepal Motor Vehicle Agreement: A Supply-chain Analysis of the Benefits to the North East Region of India." *Economic and Political Weekly, LIX*. https://www.epw.in/journal/2024/46/global-value

United Nations Office for South-South Cooperation, *India-UN Development Partnership Fund.* United Nations Office for South-South Cooperation (UNOSSC).

UN (2015). *Transforming Our World: The 2030 Agenda for Sustainable Development. A/res/70/1.* United Nations.

Unknown Authors, *Narendra Modi.* 5 May. http://www.narendramodi.in/the-south-asian-satellite-some-highlights-535325

Upadhyay, Shreya (2021). "India's Disaster Relief Initiatives." *Indian Foreign Affairs Journal*, Prints Publications Pvt. Ltd., *16*: 170. https://www.jstor.org/stable/10.2307/48720488

Xavier, Constantino (2018). *Bridging the Bay of Bengal: Towards a Stronger BIMSTEC.* Carnegie India.

4

Disaster Preparedness Perspectives in West Jaintia Hills, Meghalaya

Nambha Pohrmen and *Tiewri Lyngdoh Nonglait*

Introduction

Every year, thousands of students in disaster-prone regions face unpredictable natural and human calamities, yet many lack the necessary preparedness to respond effectively. Disasters pose a profound threat to educational institutions, not only disrupting learning but also endangering lives and straining already limited resources (United Nations Office for Disaster Risk Reduction, 2021). Schools and colleges, meant to be safe havens of knowledge and growth, can quickly turn into sites of chaos and vulnerability in the face of various disasters. In regions like Meghalaya, marked by seismic activity and frequent landslides, this vulnerability is amplified, making comprehensive disaster preparedness not just a necessity but a moral imperative.

Despite global advancements in disaster management, many educational institutions remain inadequately prepared. While policies exist, their implementation is often inconsistent, leaving students and teachers exposed to preventable risks. This study critically examines the current state of disaster awareness and preparedness among students and teachers in the West Jaintia Hills District of Meghalaya. Through this chapter, we seek to uncover existing gaps, understand institutional challenges, and propose actionable strategies to build a culture of disaster resilience in educational settings.

The importance of disaster preparedness in schools and colleges has been widely acknowledged in academic literature. Research suggests that fostering disaster resilience requires more than just theoretical awareness—it demands systematic training, resource allocation, and proactive institutional planning (Paton, 2003; Kawasaki *et al.*, 2022). Patel *et al.* (2023) emphasize that preparedness measures must go beyond policy frameworks to engage students and educators in hands-on drills and response simulations, ensuring that knowledge translates into action when disaster strikes.

However, awareness alone does not guarantee preparedness. Kawasaki *et al.* (2022) highlight that many educators possess theoretical knowledge of disaster risks but struggle to implement effective preparedness measures due to resource constraints and institutional limitations. In regions like Meghalaya, where natural disasters such as earthquakes and landslides are a constant threat, educational institutions require more than just theoretical knowledge—they need structured disaster education programmes, well-equipped infrastructure, and proactive institutional policies. Without these measures, even the best awareness campaigns may fail to translate into meaningful action during emergencies.

Furthermore, Patel *et al.* (2023) argue that disaster preparedness is not merely about response strategies—it is about fostering a mindset of readiness. Schools that regularly conduct drills and integrate disaster education into their curriculum tend to develop stronger community resilience. Their findings align with the need for a holistic approach that combines awareness, training, infrastructure development, and community engagement.

The primary objectives of the study are:

1. To assess the awareness and understanding of disasters among students and teachers in the West Jaintia Hills District, Meghalaya.
2. To examine the preparedness measures and institutional support available in schools and colleges.
3. To identify the gaps in training, infrastructure, and institutional support for disaster preparedness.
4. To propose recommendations for enhancing disaster readiness in educational institutions.

By addressing these objectives, this chapter aims to contribute to the broader discourse on disaster management in the education sector, particularly in regions with high vulnerability to natural hazards.

Research Contribution

This study builds upon existing literature by offering a focused assessment of disaster preparedness in the educational institutions of Meghalaya, specifically in the West Jaintia Hills District. By identifying key vulnerabilities and institutional shortcomings, it seeks to bridge the gap between theoretical knowledge and real-world preparedness. The findings will not only contribute to academic discourse but will also serve as a call to action-advocating for safer learning environments and empowering educational academies with the tools they need to withstand future disasters.

Methodology

Research Design

This study employs a descriptive research design to assess the awareness, perception, and preparedness levels of students, teachers, and principals regarding disaster management in educational institutions within the West Jaintia Hills District, Meghalaya. A descriptive approach was chosen as it allows for a detailed examination of existing disaster awareness, preparedness measures, knowledge gaps, and institutional readiness. This research is quantitative in nature, utilizing survey-based data collection to derive statistical insights into the preparedness levels among respondents.

Study Area and Population

The research was conducted in the Educational Institutions across the West Jaintia Hills District, Meghalaya, a region highly vulnerable to earthquakes and landslides due to its geographical and seismic characteristics. The study targeted students, teachers, and principals, as they play crucial roles in disaster preparedness. Students are among the most affected groups during disasters, while teachers and principals are responsible for implementing disaster management strategies and ensuring safety within educational institutions.

Data Collection Methods

Primary data were collected using Google Forms, which were distributed to students, teachers, and principals after obtaining permission from educational authorities. The survey included structured, close-ended questions covering:

(a) **General disaster awareness:** Knowledge of common natural and human disasters in Meghalaya.

(b) **Perception of safety:** Confidence in the current disaster preparedness levels in the educational institutions.

(c) **Preparedness measures:** Availability of emergency response plans, participation in disaster drills, and accessibility of safety resources.

(d) **Institutional support and training:** Frequency and effectiveness of disaster management training programmes.

(e) **Recommendations for improvement:** Suggestions for enhancing disaster preparedness in educational institutions.

Respondents were required to answer based on their personal experiences and institutional practices. The self-administered online survey method was chosen for its efficiency in reaching a large number of participants while minimizing disruptions to academic activities.

Data Analysis Techniques

After collection, the survey data were exported to MS Excel and SPSS for analysis. The following statistical techniques were applied:

(a) Descriptive statistics (percentages, means, standard deviations) to summarize overall trends in awareness and preparedness.

(b) Skewness and kurtosis to assess the distribution of responses and detect any significant biases in the data.

(c) Chi-square tests to examine associations between categorical variables, such as the relationship between disaster preparedness training and perceived institutional safety.

(d) Correlation coefficient (r) to determine the strength of relationships between different factors, such as the impact of training frequency on confidence in disaster response.

Ethical Considerations

To ensure ethical compliance, the study adhered to the following principles:

(a) **Informed consent:** Participants were informed about the purpose of the study, and their participation was voluntary.
(b) **Confidentiality:** All responses were anonymized, and no personally identifiable information was collected.
(c) **Institutional approval:** Permission was obtained from school and college administrators before distributing the surveys.

Results

Demographic Profile of Respondents

Demographic details of the respondents are presented in Table 4.1. The study surveyed 202 participants, including students from various high schools, higher secondary schools, universities, and colleges in West Jaintia Hills, Meghalaya. The sample consisted of 51.5 per cent females and 48.5 per cent males. The distribution of educational levels was relatively balanced, with 51 per cent of participants at the school level and 49 per cent at the college level. The majority of respondents (68.81%, n = 139) were aged 15-19 years, while the remaining 31.19 per cent (n = 63) were within the 20-25 age range.

Table 4.1: Demographic characteristics of students

Variables	Frequency (N)	Percentage
Gender		
Male	98	48.51
Female	104	51.5

(*Contd.*)

Variables	Frequency (N)	Percentage
Educational level		
School	103	51
College	99	49
Age		
15-19	139	68.81
20-25	63	31.19

The demographic details of teachers and principals are presented in Table 4.2. The study surveyed 17 participants, including teachers and principals from various educational institutions in West Jaintia Hills, Meghalaya. The sample consisted of 82.35 per cent females and 17.65 per cent males. Most of the participants were teachers (88.2%), and only 11.8 per cent principals participated. The majority had 1 to 14 years of experience and the others 14 to 28 years, providing insight into their level of exposure to disaster-related situations in educational settings.

Table 4.2: Demographic characteristics of teachers and principal

Variables	Frequency (N)	Percentage
Gender		
Male	3	17.65
Female	14	82.35
Position		
Teacher	15	88.2
Principal	2	11.8
Number of years in the current profession		
1-14	11	64.71
14-28	3	17.65

Awareness of Disasters

The levels of disaster awareness among students, teachers, and principals are shown in Figure 4.1. The types of disasters that can occur in our state are mostly aware of by all the respondents, including students, teachers, and principals. Most respondents demonstrated high awareness of disasters such as earthquakes (72.3%), landslides (16.3%) in students, as well as earthquakes (76.5%) and landslides (23.5%) in teachers and principals; however, awareness significantly decreased when it came to other types of disasters such as fire hazards and man-made hazards.

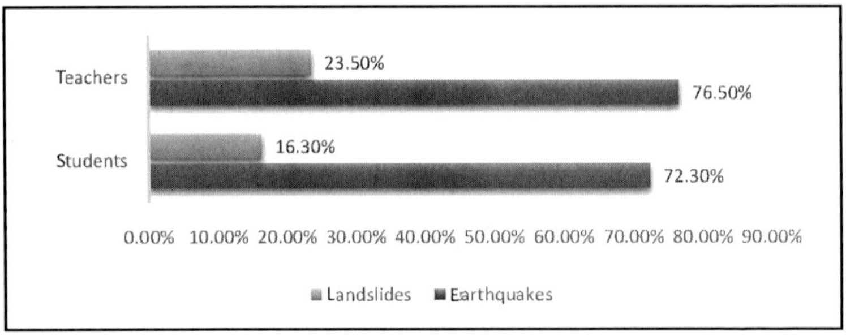

Fig. 4.1: Most recognized disaster types by respondents.

The majority of the students reported gaining information from school/college subjects and media (56.4%) and (26.7%), respectively, while a smaller percentage cited community/family discussion (12.9%), followed by workshops/seminars (1.5%), others (1.5%), and (1%) reported gaining information from all sources. Teachers reported that disaster awareness and knowledge came primarily from school/college subjects (58.8%), followed by workshops/seminars (23.5%), dedicated lessons/chapters (11.8%), and extracurricular activities (5.9%).

About 68.8 per cent of students were aware or familiar with the disaster history prevalent in the state, while 31.2 per cent were unaware. Meanwhile, the results also revealed that some of the students (6.9%) were not aware of the differences between natural disasters and man-made disasters, which could be the reason why there were many students who lacked information or knowledge about natural as well as man-made disasters. In addition, a significant number of students have never experienced a disaster personally, which indicates the level of importance for awareness and preparedness among students in various education institutions.

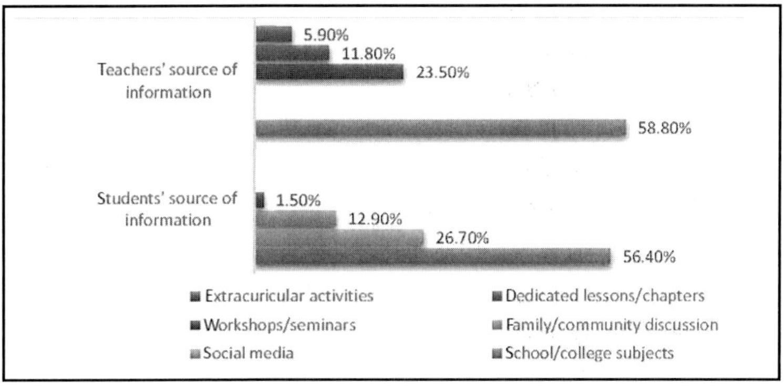

Fig. 4.2: Sources of awareness.

The results also showed that the respondents received limited training and mock drills for disaster awareness and preparedness, highlighting the critical gap between disaster awareness and preparedness within educational institutions in the West Jaintia Hills, Meghalaya. Overall, the level of awareness of disasters represented in Table 4.3 indicates that the respondents possessed a high level of disaster awareness.

Table 4.3: Disaster awareness in the respondents' perspective

Respondents	Questions	Responses	Frequency (N)	Percentage
Students	Are you aware of the types of disasters that can occur in our state?	Yes	104	51.5
		No	28	13.9
		Maybe	70	34.7
	Which of these disasters are you aware of?	Earthquakes	146	72.3
		Landslides	33	16.3
		Floods	9	4.5
		Cyclones	10	5
		Others	2	1
		All of the above	2	1

(*Contd.*)

Respondents	Questions	Responses	Fre-quency (N)	Percen-tage
	How did you learn about disaster awareness?	School/college curriculum Social media Family/ community discussions Workshops/ seminars Others All of these	114 54 26 3 3 2	56.4 26.7 12.9 1.5 1.5 1
	Are you aware or familiar with the natural disaster history that are prevalent in our state?	Yes No	139 63	68.8 31.2
	Do you know the difference between natural and man-made disasters?	Yes No	188 14	93.1 6.9
	Have you ever experie-nced a disaster person-ally?	Yes No	54 148	26.7 73.3
	Have you participated in any mock drills for disaster preparedness?	Yes No	54 148	26.7 73.3
Teachers	Are you aware of the types of disasters that are most likely to occur in our state?	Yes No	170 0	100 0
	Which disaster do you think pose the highest risk to your School/ College?	Earthquake Landslides	1 34	76.5 23.5
	Have you received any training on disas-ter awareness or prepare-dness?	Yes No	7 10	41.2 58.8
	Does your School/ College include disaster management topics in the curriculum?	Yes No	10 7	58.8 41.2

(*Contd.*)

Respondents	Questions	Responses	Frequency (N)	Percentage
	From which class onwards are disaster management topics taught?	5-10 11 above Not sure	8 2 7	47.0 611.8 41.18
	How are disaster management topics covered in the curriculum?	As part of regular subjects Workshops/ seminars Dedicated lessons/chapters. As part of curriculum and workshops	10 4 2 1	58.8 23.5 11.8 5.9

Disaster Preparedness and Readiness

The responses to questions related to the degree of preparedness and readiness of the educational institutions in West Jaintia Hills, Meghalaya, are presented in Figures 4.3 and 4.4.

Understanding emergency responses is crucial for preparing for a disaster; however, our research shows that a significant number of students lack even the most fundamental understanding in this area. When asked if they knew the emergency numbers to call in the case of a disaster, only 44.6 per cent of the students responded affirmatively, whereas 55.4 per cent reported that they did not. This indicates a gap in basic emergency awareness.

The students' responses to emergency preparedness actions differed. The most popular response was first aid basics (46.5%), followed by recognition of safe zones (35.14%). However, only 14.4 per cent knew how to evacuate and only 2 per cent were familiar with emergency kits. Alarmingly, 1 per cent of the students claimed that they had no knowledge of any preparedness actions, emphasizing the need for more education in this area.

The ability to use emergency tools, such as fire extinguishers, ropes, or first-aid kits, is a critical skill during emergencies. While 31.7 per cent of the students felt confident using these tools, 41.6 per cent felt partially confident, and 26.7 per cent reported that they did not. This suggests that while some students are familiar, many have not received hands-on training in using essential emergency equipments.

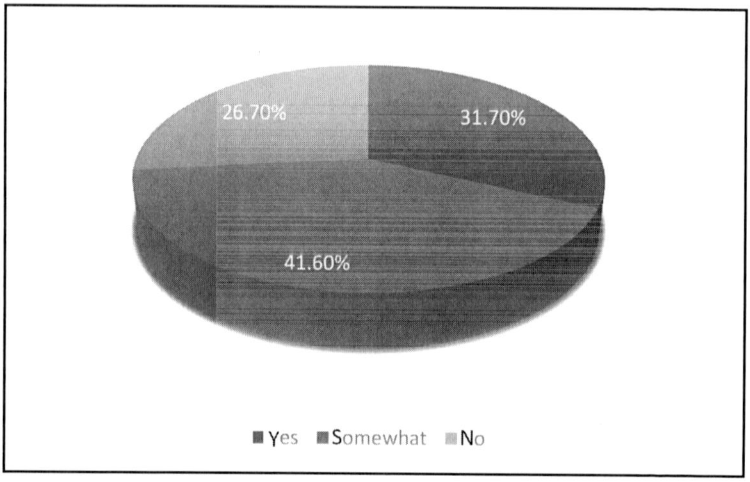

Fig. 4.3: Students' confidence in using emergency tools.

When asked about their awareness of government and community organizations involved in disaster management, only 41.1 per cent knew about government agencies, such as the National Disaster Response Force (NDRF), while 58.9 per cent did not. Correspondingly, only 35.1 per cent of the students could identify local NGOs involved in disaster management, but 64.9 per cent were unaware of their existence.

Educational institutions play an important role in disaster preparedness, yet only 38.6 per cent of the students admitted that their institution provided disaster preparedness training, whereas 61.4 per cent said no. This indicates that the majority of students had not received disaster training education at their institution. The findings reveal a significant gap in disaster preparedness drills, with 73.3 per cent of students reporting they have never participated in such drills, while only 26.7 per cent have had some exposure. This lack of regular practice may contribute to lower confidence levels in responding effectively during emergencies, highlighting the urgent need for more frequent and structured preparedness exercises in educational institutions.

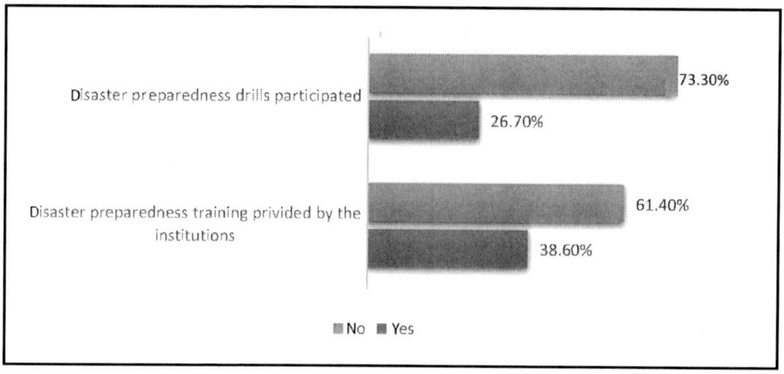

Fig. 4.4: Frequency of disaster preparedness training received.

In addition, family preparedness was insufficient. About 31.7 per cent of students admitted to having a family preparedness plan; however, the majority of students (68.3%) admitted to having none. On the positive side, 60.4 per cent had at least discussed disaster preparedness with family and friends, indicating that there was knowledge of concern about safety, although they may not have formal planning.

When asked how they would react if they were caught in a disaster at school or college, their responses were encouraging. A notable group of students (74.37%) demonstrated comprehensive disaster-response knowledge, detailing multiple safety issues. However, 25.63 per cent displayed uncertainty or reactive behaviour. This suggests that, although many students knew how to respond during a disaster, a significant portion remains unsure of how to react in such situations.

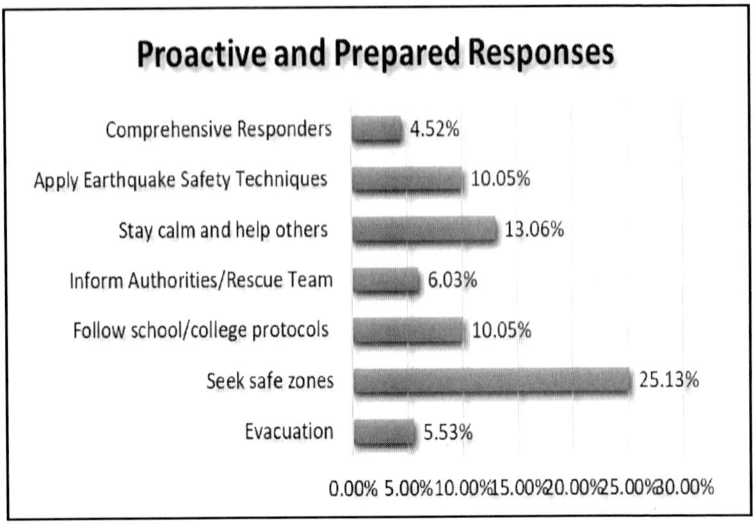

(i)

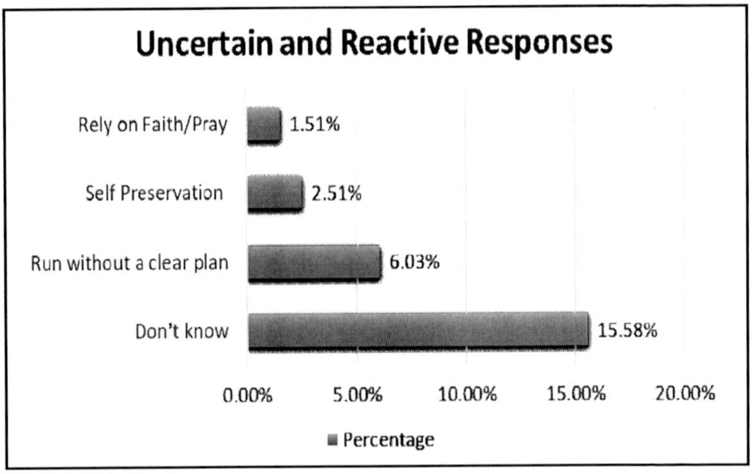

(ii)

Fig. 4.5 (i) & (ii): Students' response strategies during disasters: patterns and gaps in preparedness.

The responses of teachers and principals provide further insight into the preparedness of educational institutions. When asked if their school or college had a catastrophe management strategy, only 23.5 per cent confirmed it, 29.4 per cent denied, and 47.1 per cent were unsure. This confusion shows a lack of communication or awareness among the staff members.

On the bright side, 94.1 per cent stated that their institution had emergency tools, indicating that at least some basic supply was available. However, the mere presence of these instruments does not guarantee their efficient usage.

In terms of disaster preparedness drills, only 5.9 per cent of the institutions conducted them quarterly, 47.10 per cent conducted them annually, and 47.10 per cent reported never conducting drills. Regular drills are vital for ensuring that students and staff are adequately prepared in the case of an emergency; therefore, this is a key area for improvement.

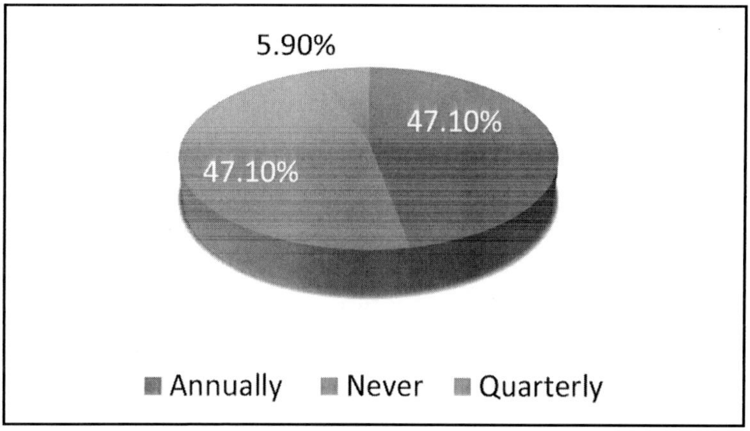

Fig. 4.6: Institutions' disaster preparedness drills frequency.

A well-marked disaster management map can help students and staff navigate disaster situations efficiently; unfortunately, only 17.60 per cent of the institutions had one. Among those that did, none reported it being fully assessable; 29.40 per cent said it was only partially visible, and 70.6 per cent admitted it was not visible at all. A disaster management committee is also vital for coordinating preparedness efforts within educational institutions. However, only 17.60 per cent of respondents stated that their institution had a designated disaster management committee, while 82.4 per cent said there was none. Without a dedicated team that oversees disaster preparedness, it is unlikely that institutions will be able to improve their readiness effectively.

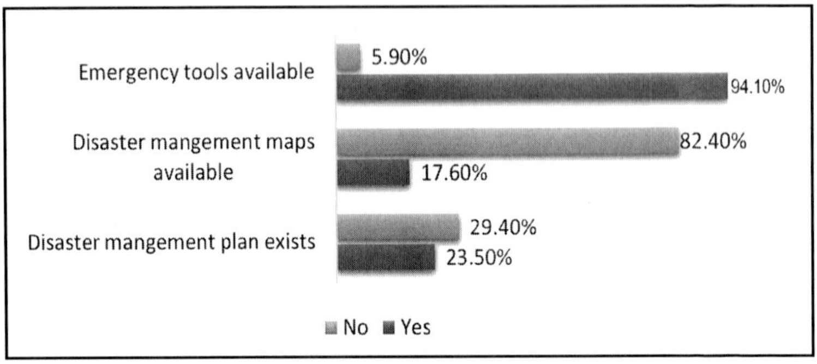

Fig. 4.7: Institutional preparedness measures.

When asked whether they believed their institution's infrastructure was disaster resilient, only 17.6 per cent said yes, while 11.85 per cent said no, leaving a large percentage unsure. This suggests that many educational institutions may not have assessed their buildings for disaster risk.

Encouragingly, most educators are eager to improve their disaster preparedness. 64.7 per cent of the respondents stated that they would like to receive additional training or resources on disaster preparedness, indicating a willingness to enhance their knowledge and contribute to a safer learning environment.

Moreover, Table 4.4 shows that the students also provided information regarding the level of preparedness and readiness in their schools and colleges, and there was a neutral response of slightly agreeing with a mean score of 3.31 (SD = 0.92). This indicates that some students believe their educational institute is somehow prepared and recognizes the effort being made towards disaster preparedness, while others believe their institutions still have room for improvement. However, when asked to rate their level of confidence in the case of a disaster, students were noticeably less confident. The mean score was 1.99 (SD = 0.67), which was interpreted as somewhat confident. Many students doubt their ability to respond effectively to emergencies, which is unsurprising since over 60 per cent have never received disaster preparedness training. Without regular drills or hands-on training, quick and informed decision making becomes challenging.

The teachers' role in ensuring students' safety is crucial during disasters. However, the results indicate that the teachers themselves do not feel fully prepared. The mean confidence level was 2.12 (SD = 0.68), which was interpreted as 'somewhat confident'. While this demonstrates that many educators believe that they could provide some guidance, it also shows a disinclination due to lack of training or clear disaster response protocols in their institutions. When assessing how prepared institutions handle disasters, the response is even more concerning. The mean score was just 1.71 (SD = 0.67), meaning that institutions are perceived as only 'somewhat prepared'. This aligns with earlier findings that many schools and colleges lack disaster-management plans, emergency maps, and designated response committees.

Table 4.4: Frequency and Percentage of Disaster Preparedness and Readiness

Respondents	**Questions**	**Responses**	**Frequency (N)**	**Percentage**
Students	Do you know what emergency members to call in case of disaster?	Yes No	90 112	44.6 55.4

(*Contd.*)

Respondents	Questions	Responses	Frequency (N)	Percentage
	What emergency preparedness action do you know?	First aid Basics	94	46.5
		Identifying safe zones	73	36.14
		Evaluation procedures	29	14.4
		Assembling emergency kits	4	2
		None	2	1
	Do you know how to use emergency tools like fire extinguishers, ropes, or first aid kits?	Yes	64	31.7
		No	54	26.7
		Somewhat	84	41.6
	Are you aware of the roles of government agencies like the National Disaster Response Force (NDRF) in disaster management?	Yes	83	41.1
		No	119	58.9
	Are there any local community organizations or NGOs in the district that you know are involved in disaster management?	Yes	71	35.1
		No	131	64.9
	Does you school/college provide disaster preparedness training?	Yes	78	38.6
		No	124	61.4
	Have you participated in any mock drills for disaster management?	Yes	54	26.7
		No	148	73.3
	Does your family have a disaster preparedness Plan?	Yes	64	31.7
		No	138	68.3
	Have you discussed disaster preparedness with your family or friends?	Yes	122	60.4
		No	80	39.6

(*Contd.*)

Respondents	Questions	Responses	Frequency (N)	Percentage
	What would you do if you were caught in a disaster at School/College?	Proactive and prepared responses Uncertain and reactive responses	150 52	74.37 25.63
Teachers	Does your disaster at School/College have disaster management plan in place?	Yes No Not sure	4 5 8	23.5 29.4 47.1
	Are there emergency tools and equipment (e.g., first aid kits, fire extinguishers) available and accessible in your institution?	Yes No	16 1	94.1 5.9
	How often are disaster preparedness drills conducted in your institution?	Annually Quarterly Never	8 1 8	47.1 5.9 47.1
	Does your School/College have a disaster management map (e.g., showing evacuation routes, safe zones, and emergency points)?	Yes No	3 14	17.6 82.4
	If yes is the disaster management map visible and accessible to all staff and students?	Partially No	5 12	29.4 70.6
	Is there a designated Disaster Management Committee in your School/College?	Yes No	3 14	17.6 82.4
	Do you think the infrstructure of your institution is disaster resilient?	Yes No Not sure	3 2 12	17.6 11.8 70.6
	Would you like to receive additional training or resources for disaster preparedness?	Yes No	11 6	64.7 35.3

Statistical Analysis

The data collected for this study were analysed using the Statistical Package for the Social Sciences (SPSS). The data were thoroughly checked to ensure the accuracy of entries before the analysis initiation. The demographic and socio-economic characteristics of the respondents are presented as percentages (%). A univariate normality test was performed to confirm a normal distribution. When appropriate, Chi-square test and Pearson correlation were used. A p-value of less than 0.05 was considered to be significant.

Students' Perspective

Table 4.5: Skewness and Kurtosis

Disaster Preparedness variables	Skewness statistics	Standard error	Kurtosis statistics	Standard error
A	1.213	.172	4.113	.343
B	.315	.172	-1.743	.343
C	1.166	.172	1.308	.343
D	.795	.172	-1.383	.343
E	3.386	.172	9.558	.343
F	-1.095	.172	-.810	.343
G	1.715	.172	2.547	.343
H	-.490	.172	-1.778	.343
I	-1.095	.172	-.810	.343
J	-.255	.172	-1.955	.343
K	-.820	.172	-1.341	.343
L	-.338	.172	-.071	.343
M	.424	.172	-1.839	.343
N	-.626	.172	-1.625	.343

(*Contd.*)

Disaster Preparedness variables	Skewness statistics	Standard error	Kurtosis statistics	Standard error
O	.374	.172	.102	.343
P	-.012	.172	-.827	.343
Q	-.184	.172	-1.596	.343
R	-.360	.172	-1.889	.343

Table 4.6. Test of significance using Chi-square

Disaster preparedness variables	Chi-square	Asymp.sig.
A	456.09	0
B	14047	0
C	296.47	0
D	26.78	0
E	146.94	0
F	45.35	0
G	238.65	0
H	11.10	0.001
I	45.35	0
J	3.14	0.076
K	28.27	0
L	162.08	0
M	8.45	0.004
N	17.49	0
O	114.24	0
P	37.43	0
Q	6.16	0.046
R	6.16	0.013

Table 4.7: Correlation matrix of different variables

	Knowledge & Awareness				Perception of Safety							Preparedness Levels						
	A	B	C	D	E	F	G	H	I	J	K	L	M	N	O	P	Q	R
A	1																	
B	-.057	1																
C	-.044	.152*	1															
D	.074	.234**	.083	1														
E	.006	.090	.093	.235**	1													
F	-.044	.149*	.025	.208**	.074	1												
G	.087	-.079	.071	.004	-.069	-.486**	1											
H	-.031	.194**	.013	.046	.054	.097	-.086	1										
I	.019	.224**	.036	.159*	.119	.219**	-.060	.427**	1									
J	.051	.083	.030	.164*	.084	.121	-.066	.079	.075	1								
K	.061	.111	-.029	.155*	-.027	.217**	-.044	.119	.217**	.304**	1							
L	.105	.056	.007	.057	-.028	.067	-.067	-.083	-.052	-.036	-.074	1						
M	-.068	.082	.132	.199**	.138	.109	-.124	.173*	.202**	.239**	.324**	-.153*	1					
N	.028	.128	.016	.049	.079	.137	-.206**	.136	.137	.263**	.141*	-.075	.168*	1				
O	.151*	.117	.055	.004	.006	-.074	.008	.323**	.113	.224**	.165*	.053	.114	.189**	1			
P	-.183**	-.022	.017	.085	-.004	.076	.003	.072	.042	.371**	.249**	-.033	.184**	.181*	.150*	1		
Q	.027	.220**	.051	.140*	-.077	.161*	-.047	.076	.161*	.207**	.178*	-.040	.139	.232**	.179*	.181*	1	
R	-.116	.248**	-.078	.175*	.071	.083	-.160*	.224**	.199**	.250**	.208**	-.144*	.220**	.217**	.098	.178*	.154*	1

*p value significance at 95% **p value significance at 98%

Teachers' Perspective

Table 4.8: Distribution characteristics: Skewness and Kurtosis

Disaster Prepared-ness Variables	Standard Error of Skewness	Skewness	Standard Error of Kurtosis	Kurtosis
A	.550	–	1.063	-.149
B	.550	-.394	1.063	-2.109
C	.550	.677	1.063	-1.766
D	.550	.394	1.063	-2.109
E	.550	.184	1.063	-1.007
F	.550	1.372	1.063	-.149
G	.550	-.496	1.063	-1.357
H	.550	.522	1.063	-.443
I	.550	–	1.063	1.665
J	.550	-.994	1.063	-1.166
K	.550	–	1.063	1.665
L	.550	–	1.063	.162
M	.550	4.123	1.063	17.000
N	.550	.161	1.063	-674
O	.550	-456	1.063	-611

Table 4.9: Test of significance using Chi-square

Variable	Chi-square	p-value
B	4.765a	0.029
C	.529a	0.467
D	1.471a	0.225
E	.529a	0.467
F	6.647b	0.248
G	4.765a	0.029
H	1.529c	0.465
I	5.765c	0.056
J	7.118a	0.008
K	2.882a	0.09
L	7.118a	0.008
M	10.706c	0.005
N	13.235a	0
O	3.294c	0.193
P	3.647c	0.161

Table 4.10: Correlation matrix of different variables

	B	C	D	E	F	G	H	I	J	K	L	M	N	O	P
B	1														
C	-.182	1													
D	-.171	-.132	1												
E	.464	.457	-.368	1											
F	.129	.338	-.214	.646**	1										
G	.308	.182	-.410	.663**	.632**	1									
H	-.010	.392	-.215	.349	.333	.182	1								
I	.544*	.023	-.519*	.375	.210	.381	.079	1							
J	.107	.240	.019	.387	.187	.257	.326	.197	1						
K	-.358	.509*	-.064	.278	.172	.054	.188	-.228	.378	1					
L	-.257	.240	-.304	.074	.081	.257	.326	.454	.190	.040	1				
M	-.158	.109	-.028	.045	.245	-.021	.177	-.037	.515*	.274	.117	1			
N	.139	.209	-.185	.299	-.444	-.139	-.073	.172	.116	.161	.116	-.171	1		
O	.314	.385	-.600*	.499*	.070	.097	.482*	.171	.148	.270	.148	.119	.413	1	
P	.662**	.190	-.511*	.528*	.185	.172	.309	.451	-.027	.091	-.027	-.188	.265	.731**	1

*p value significance at 95% **p value significance at 98%

Table 4.11: Mean and standard deviation of disaster preparedness and readiness

Respondents	Questions	Mean (M)	Standard deviation (SD)	Interpretation
Students	The level of disaster preparedness and readiness possessed by their school/college	3.31	0.92	Neutral-Agree
	Students' level of confidence to respond during a disaster	1.99	0.67	Somewhat confident
Teachers	Teachers' confidence level in guiding students during a disaster	2.12	0.68	Somewhat confident
	The degree to which educational institutions are equipped to manage a disaster	1.71	0.67	Somewhat prepared

Challenges to Disaster Preparedness in Schools and Colleges

Many educational institutions, despite having access to disaster awareness and knowledge, struggle to translate this awareness into effective preparedness. Various challenges hinder the implementation of comprehensive disaster preparedness measures within these institutions. Teachers have identified several critical gaps that limit their institute's ability to respond efficiently to crises. Below are the key challenges reported by teachers.

(i) **Inadequate infrastructure and facilities (35%):** Many educational institutions lack the necessary infrastructure to handle crisis situations, such as disaster-resistant buildings, escape signs, defined evacuation routes, and enough emergency exits. Given the region's vulnerability to natural disasters, particularly earthquakes and landslides, building safe and sustainable learning infrastructure with all the institutional amenities required to provide a secure environment for educational stakeholders is not only important, but crucial.

(ii) **Insufficient training and awareness programmes (29%):** Teachers' responses indicated that they do not receive enough training, frequent practice exercises, workshops, direction, or organized programmes to keep them informed on emergency protocols. Inadequate incorporation of actual disaster preparedness information, mock exercises, and hands-on training activities into curricula can impair an institution's capacity to manage a crisis effectively, directly endangering the lives of everyone involved. In order to combat any crisis, practical actions are crucial, even while theoretical lessons on disaster awareness, preparedness, and readiness are necessary.

(iii) **Financial constrains (24%):** Lack of finance and financial difficulties are among the factors contributing to institutions' poor preparedness and catastrophe readiness. Financial constraints make it difficult to purchase necessary supplies like training sessions, updated evacuation plans, safe infrastructure, survival packs, and so on. Teachers stress that budget constraints limit schools' capacity to improve their preparedness for disasters, highlighting the necessity of greater funding and assistance from governmental and non-governmental organizations to adequately address these shortcomings. Hence, institutions find it difficult to put required safeguards into place without enough financial assistance, thereby compromising the safety and resilience of their communities.

(iv) **Limited equipment and resources (18%):**The difficulties with scarce supplies and equipment, which severely impair catastrophe readiness in schools, colleges, and universities, were also brought to light by the teaching staff. Essential emergency supplies like fire extinguishers, first aid kits, emergency alarms, and protective gear are typically lacking in many institutions, and those that are available are frequently out-of-date or badly maintained. Access to training materials is further restricted by financial limitations, which leaves staff and students unprepared for crises.

(v) **Unpredictability of disasters (6%):** Since schools, colleges or universities cannot always predict what kind of disaster will occur or when, the unpredictability of disasters presents a significant challenge. While some disasters occur unexpectedly, leaving little to no time for preparation, others may be unprecedented in a given region, making it difficult to create effective response plans. Because of this uncertainty, it is practically hard to properly prepare for every situation that could occur, which results in gaps in disaster preparedness efforts. Schools frequently find it difficult to commit resources and training for unknown dangers, which increases the likelihood that they will not be ready for a crisis when it strikes.

Recommendations for Enhanced Readiness

Based on the qualitative findings from the respondents, key themes have been identified to enhance disaster preparedness and readiness in educational institutions.

(a) **Regular disaster drills and training programmes**: Students and faculty emphasised the need for frequent disaster training and drills to enhance their confidence in responding effectively to emergencies. Beyond personal preparedness, these initiatives would also enable them to educate their families and communities on disaster response strategies. The findings indicate that many educational institutions in the region have not prioritised disaster preparedness training and in some cases schools and colleges have never conducted disaster related programmes or drills. A key barrier to implementation appears to be financial constraints, with institutions struggling due to limited funding for such initiatives. To address this gap integrating structured disaster preparedness activities such as regular mock drills hands-on training and clearly defined evacuation procedures can significantly improve readiness among students, and faculty members. Establishing a culture of preparedness within educational institutions will not only enhance individual response capabilities but also contribute to a more resilient academic community in West Jaintia Hills.

(b) **Improved access to emergency resources**: Additionally, it is advised that institutions install, upgrade, and make sure they have functional emergency supplies including fire extinguishers, first aid kits, emergency exit maps, protective equipments and conspicuous signs indicating safe areas and evacuation routes. It is evident from the respondents' suggestions that there may have been fewer or no resources available at all, or that even if there are emergency tools, some of them are out-of-date. This is concerning because in times of crisis, this can be dangerous and endanger life; therefore, all educational institutions should set up well-equipped resources to guarantee the safety of the faculty and students collectively.

(c) **Infrastructure and safety measures**: Students and faculty emphasised the need for improved infrastructure and safety measures to enhance disaster readiness in educational institutions. A significant number of knowledge centres do not have a structurally safe, disaster resilient buildings, which compromises the safety of students and staff in the event of disaster. To address this concern, respondents recommended conducting regular safety audits, guaranteeing that the infrastructure of the centre meet disaster resilient construction standards, and installing visible emergency signage.

(d) **Community-integrated disaster management:** To strengthen partnerships between Educational Institutions, Families and Authorities, a well-organized disaster management committee may be established with participation with institutions, families, local government agencies, non-governmental organizations, and families. It is one of the crucial steps that the respondents emphasized as being necessary to guarantee a coordinated and efficient response. While family involvement goes beyond home preparedness—it enables parents to understand emergency protocols, and coordinate with institutions during crisis. Partnership with local authorities and NGOs provide training and resources, and government support is crucial for funding and infrastructure. A well-coordinated, multi-stakeholder approach enhances disaster resilience across academic and community levels.

(e) **Institutionalising Disaster Management—Establishing Departments and Integrating Awareness into Education:** To strengthen disaster preparedness institutions, respondents highlighted the need for both structural and academic reforms. The establishment of dedicated disaster management departments would ensure a systematic approach to preparedness, response, and mitigation. These departments would oversee emergency planning, conduct regular drills, train students and staff, and coordinate with local authorities and disaster response agencies. Additionally, integrating disaster awareness into curriculum would equip students with essential knowledge and practical skills, making preparedness a fundamental part of their education. Courses on risk assessment, first aid, evacuation procedures, and local disaster response mechanisms can help students increase their confidence in disaster readiness. Including disaster education across various subjects would foster a culture of preparedness from an early stage, ensuring long-term resilience within academic institutions.

Discussion

Awareness of Disasters

Our study examined the disaster preparedness levels among students as well as teachers in different high schools, higher secondary schools and colleges located in West Jaintia Hills, Meghalaya. It has been recorded that knowledge regarding prevailing disasters are well-known among the students, mostly attaining from school and college curriculum. This is also reported by Divya (2020) and Apronti (2015). According to earlier research, education is crucial for helping students become effectively prepared, and disaster preparedness ought to be a crucial component of the college curriculum (Jose and Dufrene, 2014; Loke and Fung, 2014). The training and curriculum should encompass all aspects of disaster management, including prevention, preparedness, mitigation, response, and recovery

(UNDP, 2012). The positive rate of disaster education is a positive sign that students' growing understanding of the potential to lower their risk of disaster will help counterbalance the widely held belief that disasters are uncontrollable and inevitable occurrences. This supports the theory put forth by Birkman *et al.* (2012) that catastrophe risk reduction may be achieved by a motivated and informed populace, but that knowledge and information on risks, vulnerabilities, and capacities must be gathered and shared.

Students generally understand what disasters are, but when asked about their knowledge of disaster and emergency training, only 38.60 per cent of them could elaborate and said they thought the training was enough, while 61.40 per cent said they had no training, that they did not rehearse disaster drills at school which is supposed to be a co-curricular activity and psychomotor aspect of development but 20 per cent of the learners indicated that emergency drills ranging from first aid drills, rescue drills, and evacuation drills were performed in their schools. Positively, 94.1 per cent of respondents said their institution had emergency supplies, meaning that at least some essentials were on hand. Only 5.9 per cent of the institutions conducted quarterly disaster preparedness drills, 47.1 per cent conducted annual drills, and 47.1 per cent said they never conducted any drills which are similar to the findings recorded by Olusola (2019).

According to the analysis, most of the students had not experienced a disaster or community crisis, 6.9 per cent were not aware of the differences between natural disasters and man-made disasters, this is concerning because college students are susceptible to all forms of disaster and have increased vulnerability to disasters (Tanner and Doberstein, 2015; Gadberry, 2018). According to Hellen (2006), co-curricular activities which included basic disaster awareness and disaster risk reduction such as mock drills, first aid training, training fire safety and other appropriate response skills, for example light search and rescue, swimming, evacuation and emergency shelter.

Disaster Preparedness and Readiness

Students' Perspective

Regarding the degree of readiness for disasters, most respondents are aware of the risk, and 56.4 per cent possess knowledge that has been taught in school and college curricula. Although they are aware of the risks they face, they have not taken any steps to prepare, just 26.7 per cent have taken part in simulated exercises, 44.6 per cent are aware of emergency contacts, and 31.7 per cent have a family preparedness plan. This has also been revealed by Bhat *et al.* (2017), Gadberry (2018) and Kawasaki (2022).

Students' perception of safety and confidence remain vague because 40.6 per cent have a neutral perception regarding their institution's preparedness and only 36.1 per cent agree that they are secured in the wake of a disaster, which was also reported by King *et al.* (2019). Regarding their degree of confidence in their ability to respond during a disaster, 54.5 per cent are somewhat confident and 23.3 per cent lack confidence.

The research demonstrated that many of the students are still ignorant of what to do in case of emergencies due to lack of education or training on the concept but concluded to contact appropriate quarters/agencies in case of emergencies. This is in agreement with the finding of Sujarwon *et al.* (2018), who identified that the level of school preparedness was lower compared to that of communities and government apparatus.

Students' knowledge and preparedness towards disaster management significantly correlated and were predictors that there is an emergent need to incorporate disaster training and skills. These findings are similar to the findings reported by Tanner and Doberstein (2015) and Gadberry, M. (2018).

The ratio of basic disaster training received by students is 38.6 per cent. A glaring 73.3 per cent have not participated in any mock drills, 55.4 per cent do not have any access to emergency contacts in the wake of a disaster and 68.3 per cent of students have no family preparedness plan.

Teachers' Perspectives

58.8 per cent of them had not been exposed to any type of first aid or disaster management programme. The same response was given for inclusion of Disaster Management topic in the curriculum. It reveals that there is general lack of information among school teachers regarding catastrophe management. This underscores the necessity for disaster safety education. These findings are similar to Ganpatrao (2015), who reported that disaster has substantial impact on students.

Most of the responses from the teaching staff revealed that preparedness for any disaster is moderate amongst them. They do not have a disaster management map (82.4%) or even a plan (47.1%). They also lack being active members of Disaster Management Committee in their Schools/Colleges (82.4%). They are quite ignorant on whether their institution have the infrastructure to manage any disasters. These findings corroborate with Mir *et al.* (2023), who reported that disaster preparedness in educational institutions varied based on several factors, including available resources, organizational structure, and regulatory requirements. Kapucu (2013) highlighted that a disaster management map or plan, conducting regular training and exercises, and inculcating strong community partnerships are vital for creating a disaster-resilient educational institution, well prepared to deal with any calamity.

Conclusion

This study highlights a significant gap between disaster awareness and preparedness within educational institutions in West Jaintia Hills, Meghalaya. While both students and teachers demonstrate a high level of awareness regarding common disasters such as earthquakes and landslides, this awareness does not necessarily translate into effective preparedness. Many schools and colleges lack structured disaster management plans, regular mock drills, and formal training programmes, leaving institutions vulnerable to unforeseen crises.

A critical analysis of the challenges faced by these institutions reveals that inadequate infrastructure, insufficient resources, and limited government support are key barriers to implementing comprehensive disaster preparedness measures. Without proper facilities, trained personnel, and financial backing, educational institutions struggle to establish a resilient disaster response framework, putting students and staff at risk.

To bridge this gap, it is essential to integrate disaster management education into school and college curricula, ensuring that students and teachers acquire the necessary skills and knowledge to respond effectively to emergencies. Additionally, conducting regular and realistic drills, enhancing institutional preparedness strategies, and investing in robust infrastructure will significantly strengthen the disaster resilience of these institutions.

By addressing these critical areas, institutions will be better equipped to manage future disasters, minimizing risks, enhancing safety, and fostering a culture of preparedness among students and staff. Strengthening institutional resilience is not just a necessity but a responsibility, ensuring that educational institutions remain safe and well-prepared in the face of potential disasters.

REFERENCES

Aftin, K.D. & John, A.O. (2016). "Level of Disaster Preparedness and Policy Implementation in Public Secondary Schools in Rhamu Town, Madera County, Kenya". *IOSR Journal of Research & Method in Education*, *6*(1): 06-11.

Alshakka *et al.* (2022). "Knowledge, Attitude, and Readiness to Practice Disaster Preparedness among University Health Students: Insights from Resource-Poor Setting". *Journal of Hunan University Natural Sciences, 49*(1): 170-178. https://doi.org/10.55463/issn.1674-2974.49.1.22

Apronti, P., Osamu, S., Otsuki, K. & Kranjac-Berisavljevic, G. (2015). "Education for Disaster Risk Reduction (DRR): Linking Theory with Practice in Ghana's Basic Schools". *Sustainability*, *7*(7): 9160-9186.

Gadberry, M. (2018). "Emergency Preparedness Among College Students at the University of Arkansas". *The Eleanor Mann School of Nursing Undergraduate Honors Theses.* Retrieved from https://scholarworks.uark.edu/nursuht/73

Hellen Mamosegare Mamogale (2011). "Assessing Disaster Preparedness of Learners and Educators in Soshanguve North schools". Unpublished Project, 12-13.

Joshi, Sonopant Ganpatrao (2014). "Knowledge and Practices of School Teachers Regarding Disaster Management." *International Journal of Health System and Disaster Management, 2*(2): 98-101. https://doi.org/10.4103/2347-9019.139055

Jose, M.M. and Dufrene, C. (2014). "Educational Competencies and Technologies for Disaster Preparedness in Undergraduate Nursing Education: An Integrative Review". *Nurse Education Today*, *34*(4): 543-551. http://dx.doi.org/10.1016/j.nedt.2013.07.021[23]

Kawasaki, Hiromi & Yamasaki, Satoko & Yamakido, Mio & Murata, Yoshihiro (2022). "Introductory Disaster Training for Aspiring Teachers: A Pilot Study". *Sustainability*, *14*: 3492. 10.3390/su14063492

Kapucu, N. & Khosa, S. (2013). "Disaster Resiliency and Culture of Preparedness for University and College Campuses". *Administration & Society*, *45*(1): 3-37. https://doi.org/10.1177/0095399712471626

King, J., Edwards, N., Watling, H. & Hair, S. (2019). "Barriers to Disability-inclusive Disaster Management in the Solomon Islands:

Perspectives of People with Disability". *International Journal of Disaster Risk Reduction, 34*: 459-466. https://doi.org/10.1016/j.ijdrr.2018.12.017

Loke, A. and Fung, O. (2014). "Nurses' Competencies in Disaster Nursing: Implications for Curriculum Development and Public Health". *International Journal of Environmental Research and Public Health, 11*(3): 3289-3303. http://dx.doi.org/10.3390/ijerph110303289

Margaret Divya, A. (2020). "An Empirical Study on Awareness of Disaster Management among Students and Staff of Various Colleges/Schools (With Special Reference to Madurai City)". https://orcid.org/0000-0001-9675-923

Matunhay (2022). "Disaster Preparedness and Sensitivity Level among Higher Education Institution Students." *International Journal of Disaster Management*. https://doi.org/10.24815/ijdm.v5i2.27150

Mir, Sumira, Shamim Ahmad Shah and Mohmmad Shafi Bhat (2023). "Disasters affecting education across time and space—a perspective from Kashmir." *Environment, Development and Sustainability*: 1-19.

Olusola, Johnson (2019). *Evaluating Disaster Preparedness among University Learners: A Study of Ekiti State University, Ado Ekiti, Nigeria*.

Patel, Ronik Ketankumar *et al.* (2023). "Disaster Preparedness and Awareness among University Students: A Structural Equation Analysis." *International Journal of Environmental Research and Public Health, 20*(5): 4447, doi:10.3390/ijerph20054447

Paton, Douglas (2003). "Disaster Preparedness: A Social-cognitive Perspective." *Disaster Prevention and Management, 12*(3): 210-216.

Rofiah, Nurul H., Norimune Kawai, and Elli Nur Hayati (2021). "Key Elements of Disaster Mitigation Education in Inclusive School Setting in the Indonesian Context." *Jàmbá: Journal of Disaster Risk Studies, 13*(1): a1159. https://doi.org/10.4102/jamba.v13i1.1159

Sujarwon, Noorhamdani & Mukhamad Fathoni (2018). "Disaster Risk Reduction in Schools: The Relationship of Knowledge and Attitudes Towards Preparedness from Elementary School Students in School-based Disaster Preparedness in the Mentawai Islands, Indonesia". *Cambridge Journals, 33*(6): 581-586.

Tanner, A. & Doberstein, B. (2015). "Emergency Preparedness Amongst University Students". *International Journal of Disaster Risk Reduction, 13*: 409-413. doi: 10.1016/j.ijdrr.2015.08.007

United Nations Development Programme (2012). *Volunteerism and Disasters. An Extract from the 2011 State of the World's Volunteerism Report*. UN Volunteers, New York.

United Nations Office for Disaster Risk Reduction (UNDRR). 2021. *UNDRR Annual Report 2021*.

Victoria, A.J. (2011). *Disaster Preparedness Education in Schools: Recommendations for New Zealand and the United States*. Published by Fulbright New Zealand.

Widowati, Evi, Wahyudi Istiono, and Adi Heru Husodo (2021). "The Development of Disaster Preparedness and Safety School Model: A Confirmatory Factor Analysis." *International Journal of Disaster Risk Reduction, 53*: 102004. https://doi.org/10.1016/j.ijdrr.2020.102004

Section II
Indigenous Practices and Traditional Knowledge

5

Environment, Nature and Society

Where Do We Go From Here?

Rekha M. Shangpliang

Introduction

The historic occasion that took place 53 years ago on 5th June, 1972, at Stockholm also known as the Stockholm Conference was the 1st International Conference that brought Environmental issues into a global platform where 20 countries and heads of states, scientists, academicians, environmentalists came together and dwelled upon the theme "Only One Earth". It is indeed amazing that 53 years down the line, we are still addressing and tackling global environmental crisis, the so-called triple planetary crisis of climate change, pollution and waste and nature and bio-diversity loss.

We are now living in a world where it can hardly be overemphasized that environment is one of the greatest concerns to every society. Even historians have not found environment easy to handle either as a concept or as a historical influence. Nevertheless, the ideas of environment and nature have played a major role both in the processes of history and its interpretation. Environment for some represents an often harrowing array of physical elements like dark gloomy forests, climate and disease, which have helped to shape the course of history.

But what is environmental history all about? Well this is a huge question that may be examined from different perspectives—whether we start from the Book of Genesis in the *Holy Bible*, or to the Greek muse of history Clio, or closer home we have various legends and folktales or so-called Creation Myths that talk about the creation of the earth-like among the Khasi tribe of Meghalaya the creation mythology or legend of U Lum Sohpetbneng hold good even today. According to this legend, the Khasi race originated from a hill *Lum Sohpetbneng* (The navel of heaven) connected to heaven through a golden ladder (*Jingkiengksiar*). This legend says that once there was a huge gigantic tree on the hill Sohpetbneng whose branches reached heaven and when the 16 (sixteen) huts or clans of the Khasi lived together in harmony (*Khathynriew trep khathynriew skum*). One day as the seven of the families were busy doing their work on the earth, the ladder was suddenly removed by the act of God's will and thus the seven families remained on earth for ever. Hence, the Khasis called themselves the seven huts or '*Hynniewtrep-Hynniewskum*'.

The French naturalist Georges Louis Leclerc in 1779 argued about the 7 grand epochs in the history of the earth—He says that in the 7 grand epochs in the history of the earth, the first 6 epochs follow closely with the biblical history of creation: ...and then comes the 7th and the last epoch commencing with the invention of agriculture when the tables begin to turn and humans assume more and more command over the earth destined to become lords of the earth. Buffon believed that this final stage when man would have dominance over the earth has already arrived in full glory. He writes:

> "The state in which we see nature today is as much our work as it is hers. We have learned to temper her, to modify her, to fit her according to our needs and desires. We have made, cultivated, fertilized the earth. Its appearance, as we see it today, is thus quite different than it was in the times prior to the inventions of the arts."

Much ink has been spilled over this so-called human-induced environmental change, the terror of the scientific revolution, the plundering of the earth so much so that we have turned the earth into "A shrinking Island in space". The ever-increasing human population depends on the environment for resource mobilization, livelihood and existence. We have drastically changed the way we look at 'mother earth'—far from living in harmony with nature we have now reached an age of the triumphant epoch of humans overpowering nature. Having said all this surely we understand that the most significant issue facing the human species in the present century and beyond is our ecological predicament—How can we survive as a species without undermining or degrading the planet Earth and its fabric of life—the very means of our survival? What is the future of rural livelihoods that depend on land and forest resources? Do we recognize the role of tribal women and sustainability? Where do we go from here?

Socio-cultural Basis of Ecology among the Khasi

The Khasi live in profound communion with nature. Nature is the pivot around which their economic, social, cultural and religious activities revolve. They sanctify nature and all its elements as gifts from God (*U Blei Nongbuh Nongthaw*) and that nature's bounties are conferred upon humanity as God's precious gifts. The earth is idolized as 'Mei *Ramew*', the terms '*Mei*' meaning mother and '*Ramew*' meaning the earth (Bareh, 2001). Thus, mother earth is an all-embracing entity comprising the land, forest, streams, rivers, hills and everything contained in it.

The Khasi Hills, the homeland of the Khasis, has wide ranges of hills of moderate altitudes, lofty plateaus, fertile valleys, rivers and streams and patches of plain borders. It is abounding in rich forests with luxuriant and lush evergreen trees of stateliest growth and extraordinary size and height. The district enjoys salubrious climate and abundant rainfall spreading almost half of the year. A great variety of minor forests like bamboo, small timber, woodlots, thatch and grass occur extensively besides many varieties of valuable timber. The climate ranges from temperate to tropical, which enables the growth of almost all varieties of crops, fruit trees and medicinal plants. The forest is their homeland and a storehouse from where they derive everything for their daily need like food, shelter, fodder, and medicine. For a Khasi, the forest is well loved home, a game sanctuary and an abode of worship around which his social, cultural and religious activities revolve. He loves a day out in the woods for hunting or fishing or just for trapping birds. For a Khasi, the forests are not merely wooded lands but they are the abodes of gods and spirits and sacred places of worship. They believe in the forces of nature like rocks, rivers, trees and shrubs as potential power and propitiate them for blessings.

Thus, the whole concept of man-nature relationship between the Khasi and their environment can be summarized in the following words of a noted Khasi writer H.O. Mawrie, "*U Khasi U im bad ka mariang bad ka mariang ka im bad u*", which when translated means "A Khasi lives with nature and nature lives with him" (Mawrie,1987).

Khasi Concept of Cosmology

The Khasi concept of Cosmology, which relates to the creation of the sun, the moon, water and air, is grounded on the regeneration and sustenance of the eco-system on earth. According to a Khasi legend in the beginning God created *Ka Ramew* (mother earth) and her husband called *Basa.* The couple lived in perfect peace and love but had no offspring. They prayed to God to bless them with children and finally, God granted them their wish and gave them 5 (five) children—they were *Ka Sngi* (sun), *U Bnai* (moon), *ka Um* (water), *Ka lyer* (air) and *ka ding* (fire). So, with the combined effort of the sun, moon, water, air, and fire, the vegetation flourished and the earth was bountiful.

The Sacred Groves or Law Kyntang: An Ecological Legacy

In Khasi life and philosophy forest has got a dignified place which is the reason why Khasi culture is animated with the significance of forest. Every village sets apart a plot of land as a sacred grove (*Law Kyntang, Law Lyngdoh* or *Law Niam*). It is a taboo and religious crime to cut down a tree or even pick up a flower or fruit from the sacred grove. The Khasi looks up to the sacred forest with awe and reverence and dare not to cast a shadow upon it. They believe that the guardian spirit of '*U Ryngkew U Basa*' dwells in these groves and their appeasement brings welfare to the people so any act of their desecration entails woes and miseries. Sacred groves are of three types (Syngai, 1999)—in a village administered by the Lyngdohs or the Khasi priests the sacred grove is called Law Lyngdoh, in the village where traditional faith or '*Ka Niam Khasi*' is still prevalent it is called *Law Niam* and if the village headman is the head of the village matters such a forest it is called *Law Kyntang.* Occasionally rites and rituals are conducted within the vicinity of the forest where the villagers pay homage to the forest deity and forest spirit. The priest along with the members of the village *durbar* perform the rituals for the peace and tranquility of the village as a whole. This being the cultural significance of the forest from the secular point of view, the sacred forest can thus be seen as a traditional and time-honoured method of forest conservation since time immemorial. This also makes these forests an integral part of the Khasi religio-cultural system and has received legal status from the Khasi and Jaintia Hills Autonomous District Councils under the Khasi and Jaintia Hills (Management and Control of Forest) Act, 1958. These forests are abounded with rich bio-diversity and are home to a number of endangered flora, fauna and plants as well as herbs of great medicinal values.

Nature in Khasi Folklore

Forest myths and legends have played an important role in enriching the ecological history of societies around the world. Such belief is manifested through many Khasi legends and folktales, which are symbolic of their respect for natural objects such as trees and animals. These Khasi folktales and legends (Khanatang bad Puriskam) are mostly woven around the various forces of nature such as hills and rocks, caves, flora and fauna. These elements of nature are personified in the legends as the mother, son, husband, wife and friend according to their natural behaviour, such as love, hate, jealousy, pride and vanity, which are projected in order to depict moral lessons.

The deep sense of eco-spirituality of the Khasi has no doubt established a close communion with God. However, with the change of time and tide over the last couple of centuries this age-old relationship that existed between God, nature and humans has been strained. The advent of the British may be seen as a watershed in the ecological history of Khasi and Jaintia Hills. The introduction of State control over forests systematically eroded people's traditional rights over natural resources posing a threat not only to their livelihood but also to the delicate balance between tribals and forest. The British looked at the forest as a revenue yielding property, which could be commercially exploited. Furthermore, the enactment of various Forest Acts and Laws for large areas of forest cover were converted into Reserved Forests thereby restricting the rights to enter or use the forest products. The situation in post-colonial era was no better. Many of the old trends such as infringement of land and forest that belonged to the local people, continued unabated.

Let us now very briefly focus on the key issues concerning the rural communities of North East India with special reference to the Khasi.

Let us first start with one of the major stumbling blocks on the road to sustainable use of natural resources, which is infringement of land and forest. There has been a large-scale transformation of landscapes inducted by the rapid economic growth and commercialization of land. Human interventions on natural environment have redefined inhabited spaces as well as identities. Such interventions have often led to disputes between groups and communities over their territories. The state of Meghalaya has an estimated forest cover of about 70 per cent of the total geographical area, i.e. 16,839 sq. km out of a total of 22,429 sq.km (2003 FSI Report).

The advent of the British colonial rule may be seen as a watershed in the ecological history of Khasi and Jaintia Hills. The rights and privileges over the natural forest and environment enjoyed by the Khasis in the past centuries were taken away through the enactment of forest laws and regulations and for the first time in the history of tribal land, there was the introduction of statutory interference on the traditional usage of the free gifts of nature.

It is a well-established fact that State control over forests has systematically eroded people's traditional rights over natural resources thus posing a threat not only to their livelihood but also to the delicate balance between tribals and the forest. In my recent study conducted in a village Lawbyrwa in Ri-Bhoi district this situation has largely been the outcome of the following two factors:

(i) Forest for Revenue, and
(ii) Forest for Development.

Forest for Revenue

Ever since the British created State forests and scientifically organised them as a profit making enterprise there has been a limitation on community control over forests. Going back to The Indian Forest Act, 1865, it sought to create a class of Government protected forest. Thereafter, the Indian Forest Act, 1878 was passed where two categories of state owned forests—'Reserved Forest' and 'Protected Forest' came into existence, with the prime intention of supplying timber to industrial sectors and to earn revenue for the State. Forests were since then seen as a revenue-yielding property, which could be commercially exploited. Obviously, the objectives of these Acts were to define the legal status and extent of proprietary rights of the Government in the forest.

During this long period of British rule, a considerable area of forested land of Khasi and Jaintia Hills, which were clad with rich forest and natural vegetation, were brought under legal ownership of the British Government, depriving the tribal forest dwellers of their natural rights. According to an estimate, the land converted into Reserved and Protected forests in Khasi and Jaintia Hills amounted to 427.79 sq. km, which is certainly a large area for a small hill district. An area of 590 acres was based in 1871-72 at the annual rent of Rs.150 from the hills situated around Shillong for the purpose of timber plantation. According to record available with the Government of Meghalaya, in 1885 Riat Laban and Shyrwat forests were constituted into Reserved Forests, in 1889 Short Round, Upper Shillong and Laitkor Green Block were created as Protected Forests and in 1892

Riatkhwan was constituted into Reserved Forests. Another two big areas were converted into Reserved Forest from Jaintia Hills, namely Saipung in 1887 and Narpuh in 1909. For an average Khasi farmer who depended heavily on *Jhum* cultivation as his livelihood and collected his daily necessities of life from the forests, the conversion of these vast natural resources into state property not only told upon their tribal economy but also on their tradition and cultural heritage.

The Reserved Forests created by British Government through the process of law was no forest in the true sense of the tribal concept. The people could see no meaning in the so-called forest where they could not enter and cut down a branch of a tree and could not utilise its produce. For a tribal, the land and the forest are free gifts of nature and they belong to the community for the common benefit of all. It was, therefore, natural on the part of the indigenous people to feel resented against the conversion of their tribal land and forests into Reserved and Protected forests restricting the use of its flora and fauna. Although the illiterate and innocent villagers could not raise any organised resistance against this injustice meted out to them, some sporadic uprising of the people was not at all lacking. The voice raised by one lakh Garos in February 1906 against reservation of forest in Garo Hills under the leadership of Sonaram Sangma is a case in point. In their petition they claimed that "their ancestors had enjoyed the privilege of living in forest in Garo Hills, made use of the forest products, sold them for their benefits besides cultivating and having rights of pastorage" (Sinha, 1993: 45).

Therefore, there seems to be a paradox between the community managed forest cover which has historically been under the people and the so-called 'scientific' management of forest by the Government.

Forest for Development

The process of economic development that followed successive implementation of various development programmes during the post-Independence period brought about another wave of pressure on the land and natural resources of the country. Large-scale economic growth, mushroom growth of industries, increase in population, urbanisation, infrastructural development through expansion of roads and rails, buildings and parks etc. resulted in the increased demand for raw materials like timber, bamboo, coal, limestone and other forest products. The forest situation during post-Independence period is no better. Immediately after Independence, the Government of India adopted the National Forest Policy in 1952, which envisaged expansion of forest cover to one-third of the country's geographical area and in case of hill states and territories to 60 per cent of the total area of the state or territory. This policy gave a new thrust to the colonial forest expansionists policy and brought additional areas of land under forest cover through implementation of various schemes under Five Year Plans such as Social Forestry programme, afforestation drive, artificial regeneration programme etc. The National Commission on Agriculture, 1976 recommended establishment of Forest Development Corporations in all states in order to undertake plantation activities and to carry out felling, processing and marketing of timber. The Commission also recommended drastic restrictions in people's access to forest and forest produce.

Our planners found in the hills, tremendous potentialities for development of various mineral and forest-based industries with inexhaustible strove of natural wealth. Numerous saw mills, plywood factories, cement factories, coal mines, carpentry and cane and bamboo workshops and paper mills grew up in and around Meghalaya. The course of development that took place during the next few decades left a deep scar on the green landscape of Meghalaya and almost dismantled its eco-system.

The diarchical form of control and management of forest in Meghalaya by the State Forest Department and the Autonomous District Councils has contributed to dual forest policies, lack of coordination and division of accountability. Both the authorities formulate their own forest policies and follow a separate set of forest rules and regulations in their respective areas of operation. The main reason for such a diarchy is that, while the State Forest Department maintains a well-qualified team of skilled technical staff, it has a comparatively meagre area of 1127.23 sq.km of forest under its jurisdiction (Statistical details given in Appendix III). The District Council on the other hand, has a large forest cover of 6250.68 sq.km under its management and control and yet it suffers from inadequate technical manpower, which has resulted in the lack of enforcement of rules.

Forest being the principal source of revenue, the District Councils could not afford to stop forest operation under prevailing fund constraints. The private forest owners who own substantial areas of forests indulged freely in sale of timber from their forests. According to the United Khasi and Jaintia Hills Autonomous District (Management and Control of Forest) Act, 1958, the private forests in Meghalaya are under the control and supervision of the District Council Authorities, but in reality, the private owners think themselves free to operate in these forests. Larger quantity of timber than what Meghalaya consumed found their ways to Assam and other states. The Acts and legislation passed by the State for conservation of forests, protection of catchment areas, and prohibition of felling and removal of timber etc. had very little effect until the Supreme Court intervened with a blanket ban on logging timber in December 1996.

Looking closely at the above two perspectives of forest utilization, viz. Forest for revenue and Forest for development, which have led to infringement of people's traditional rights over forest, it is seen that they have had a serious impact on people's livelihood thereby conflicting on their traditional rights over forest resources.

According to the Khasi Land Tenure System, there are two categories of land, namely *Ri Raid* or Community land and *Ri-Kynti* or Private land. The former category of land is set aside for the use of the community over which each Khasi member of the village has the right of use and occupancy. However, he has no right to inheritance or transfer of the land. His occupancy right lasts as long as he uses the land for himself and his family, but his right terminates if he fails to utilise the said land for three consecutive years and the land reverts to the community. However, if he has made any permanent improvement over the land or raises any structure or permanent fencing around the plot of land it becomes his permanent property as *Ri Kynti* land. This loophole present in the Khasi land tenure system offers an incentive to many people to convert *Ri Raid* land into their private property as *Ri Kynti* land. Persons with wisdom and ability could through application of labour and capital establish durable assets or *Ri Raid* land and claim the same as their individual possession at the cost of the community. This process of converting community land into private land has come a long way taking the advantage of Government grant-in-aid and loans under various development schemes like social forestry, fishery and farming. Today vast areas of *Ri Raid* land all over Khasi Hills have been transformed into farmlands, fishery ponds, orchards, forestland etc. The socio-economic implication of this trend is all too apparent, as it has led to gradual diminishing of cultivable community land and vegetations where the people had easy access to their necessities of life.

Thus, it is apparent that, "Land" and "Forest" being the natural productive resources at the base, there is an emerging conflict between the State and the people over the rights to land and its natural products ever since the advent of colonial rule in India to the present time.

The conversion of natural forests and wastelands into Reserved Forest, Protected and Unclassed Forests by the State and Autonomous District Councils led to increased restrictions on tribal access to forests, rivers and fisheries etc. The plethora of forest laws and regulations enacted by the authorities in the name of development of forest and environment severely infringed the tribal rights to natural forests and forest produces.

Conclusion

What we see today in Meghalaya is a rather contrasting situation where there is a clash between indigeneity and modernity (a term which is very relevant to Sociological discourse).While on the one hand we have indigeneity which emphasizes the collective rights of the community over land and forest resources and on the other is modernity wherein we see the State with its economist-metropolitan vision towards economic growth based on extraction and intrusion into community lands. The creation of the Autonomous District Councils in the 1950s towards protection of tribal lands was instrumental in continuation of the colonial policy of treating tribal lands as a protected enclave and it has remained so in many years even post-Independence. However, their compliance to the 'modernization' project of the State was visible in many ways (financial dependence on the State and Central Government). It is not surprising to see the ADCs becoming governmental functionaries that goes in line with State version (pan-Indian) of development that equates development with industrialization.

It is time we ponder over the same question as in the title of this chapter—Where do we go from here? Ever since our state was carved out there was some hope on the indigeneity discourse but the pedagogy of development, the road map followed the pan-Indian path which saw our lands as lagging behind and thus the need for a conventional development strategy and a new representation of forest from 'subsistence' to 'commercial'. At times, it also feels like there is a hidden hegemonic influence in many of the signed projects and development initiatives run by the Government since the prime movers may be the people in power themselves. Therefore, the picture is very grim—it leaves us to an uncertainty for our natural reserves, which are so precious. This is in line with the vision document of the Meghalaya Indigenous People's Forum's assertion that "The State should not adopt industrialization for the sake of industrialization as land to the tribal is more precious than coal and lime stones, we should develop our state according to our own genius (*The Shillong Times*, 2006). However, what a recent document in the Department of Industries Meghalaya stated, "With its rich and vast minerals, water and forest resources Meghalaya offers tremendous opportunities for investment. Therefore, we can understand that development in this context means further extraction of the untapped resources including our forests.

To conclude, while both the Central and regional planning authorities have acknowledged the specificity of North East India, they have failed to adopt a suitable alternative strategy, which could develop the region without disturbing its ecological balance and identity of the people.

REFERENCES

Arnold, David (1996). *The Problem of Nature: Environment,Culture and European Expansion.* London: Lighting Source Inc.

Bareh, Hamlet (2001). *The Economy of Meghalaya: Tradition to Transition*, Guwahati : Spectrum Publications.

Gurdon, P.R.T. (1975). *The Khasis.* Delhi: Cosmo Publications.

Kroeber, A.L and Kluckhohn Clyde (1952). *Culture: A Critical Review of Concept and Definitions.* Harvard University Peabody Museum of American Archaeology and Ethnology.

Laitflang, W.R. (1996). *Katto Katne Shaphang Ki Katto Katne*. Shillong: Andy Press.

Majumdar, R.C. (1996). *An Advanced History of India*. Delhi: Macmillan India Limited.

Mawrie, H.O. (1988). "Ka Pyrkhat U Khasi", in Miri, Sujata (ed), *An Introduction to the Study of Tribal Religions and the Khasi Milieu*, pp. 45-47. New Delhi: Concept Publishing Company Pvt. Ltd.

Mawrie, L. Barnes (2001). *The Khasi and their Natural Environment.* Shillong: Vendrame Institute Publications.

Syiem, D. *et al.* (1999). "Medicinal Plants and Herbal Medicine: A Case Study in Meghalaya". Shillong: North Eastern Biodiversity Research Cell, NEHU.

Syngai, D. (1999). "Sacred Groves of Meghalaya", in Kharbuli, B. *et al* (ed.). *Biodiversity: North East India Perspectives,* pp. 31-32, Shillong: North Eastern Biodiversity Research Cell, NEHU.

6

Knowledge, Awareness and Attitudes towards Sacred Groves among Students of Shillong College, Shillong, Meghalaya, India

An Exploratory Study

Bennathaniel H. Diengdoh

Introduction

Various types of community natural resource management exist, of which community forest management constitutes an important form. Within this umbrella category of natural resource management, sacred groves occupy a unique space in that they represent a harmonious manifestation of the religious, social and cultural value of biodiversity (Ormsby and Bhagwat, 2010). Sacred groves may be regarded as an expression of the socio-cultural and religious beliefs and practices of a community in the natural physical environment. As such, it may be argued that the extent to which they are preserved, and by extension, their integrity is contingent to varying extents upon the views, attitudes and perceptions of communities that have instituted them. As beliefs, perceptions and attitudes undergo change with time, this change may be reflected in how sacred groves are managed (Tiwari, Barik and Tripathi, 1998; Ormsby and Bhagwat, 2010; Ormsby, 2013; Diengdoh and Jayakumar, 2021; Diengdoh, 2022).

Sacred groves are generally understood to be patches of forest that are considered to be of value and are thus protected because they are considered to be of importance according to the worldview within which they have been instituted (Chandrakanth and Romm, 1991). Often, sacred groves are so regarded and accorded protection because they are believed to be the dwelling places, or places dedicated to deities and/or ancestral spirits (Gadgil and Vartak, 1976; Dorm-Adzobu *et al.*, 1991; Wessing, 1999; Byers, Cunliffe and Hudak, 2001; McWilliam, 2001; Wadley and Colfer, 2004; Khan, Khumbongmayum and Tripathi, 2008; Sarfo-Mensah *et al.*, 2010; Ormsby, 2013). Such groves have been found in Africa (Dorm-Adzobu *et al.*, 1991; Sarfo-Mensha *et al.*, 2010; Cardelús*et al.*, 2013; 2017; 2019), in Italy (Frascaroli, 2013) and throughout India as well (see Malhotra *et al.*, 2001; Khan, Khumbongmayum and Tripathi, 2008). Some enclose the tombs of saints (Deil *et al.*, 2005), while others encircle places of worship (Cardelús *et al.*, 2013; 2017; 2019; Chandrashekhara, 2011a). It has also been found that, in some groves, it is the presence of natural objects, man-made artifacts or edifices which impart the element of sacredness (Dash and Chauhan, 2002; Dash, 2005). Sacred groves also demonstrate great variation in terms of size and extent, with sites ranging less than 1 ha (Ormsby, 2013; Gupta and Sharma, 2014) to some large ones occupying several square kilometres (Dash, 2005).

Ensconced in and espoused by a range of worldviews, including Hinduism (Ormsby, 2013), Christianity (Cardelus *et al.*, 2013; 2017; 2019; Frascaroli, 2013), Islam (Deil *et al.*,

2005) and indigenous religions or belief systems (Ormsby, 2013; Diengdoh and Jayakumar, 2021; Diengdoh, 2022), sacred groves have been regarded as constituting an integral part of what may be described as "shadow" conservation networks, the existence of which is often contingent upon the beliefs, views and attitudes of the communities who have instituted them (Decher, 1997). As such, a change in the belief systems of said communities may result in a weakening of the protection regimes instituted to preserve sacred groves, which may result in their decline, a trend which has been observed by past studies (Tiwari, Barik and Tripathi, 1998; Ormsby and Bhagwat, 2010; Ormsby, 2013). Furthermore, sacred groves may be regarded as sites of socio-cultural significance, which may be promoted as important tourist destinations (Diengdoh and Jayakumar, 2021; Diengdoh, 2022), by word of mouth, television or through various social media platforms, many of which have a user base which includes numerous young individuals. Since attitudes and perceptions exert a notable influence on views and practices with regard to conservation (Anderson *et al.*, 2005), it may be argued that greater awareness and a positive attitude towards sacred groves among young people may help to garner publicity and facilitate the promotion of said groves as sites for ecotourism and/or conservation.

While past studies in sacred groves have focused on various aspects such as assessments of biodiversity (Tripathi, 2002; Upadhaya, 2002, Mishra *et al.*, 2004; Jyothi and Nameer, 2015), only a few studies have sought to understand the role played by human attitudes and beliefs (Ormsby, 2012; 2013; Diengdoh and Jayakumar, 2021) and to assess both biodiversity and human dimensions (Tiwari, Barik and Tripathi, 1998; Mohanta *et al.*, 2009; Diengdoh, 2022). In the above studies, the attitudes of communities in the immediate vicinity of the groves were studied, however, the views and perceptions of young individuals, particularly students, was not studied.

The objective of this study is, therefore, to ascertain and study the beliefs, attitudes and perceptions of college students towards sacred groves with a view of gaining an understanding of the status of sacred groves among said individuals.

Materials and Methods

Study Area

The sample population selected for the study were undergraduate and post-graduate students enrolled in various streams, and in various phases of their course of study at Shillong College, Shillong, East Khasi Hills District, Meghalaya. Shillong College, located in Laitumkhrah, Shillong (25.5728° N lat. and 91.8930° E long.) is affiliated to North-Eastern Hill University, Shillong, and offers various courses of study in, *inter alia*, Arts, Science, Commerce with a student population of 3776 enrolled in various undergraduate or post-graduate courses during the year 2023-24.

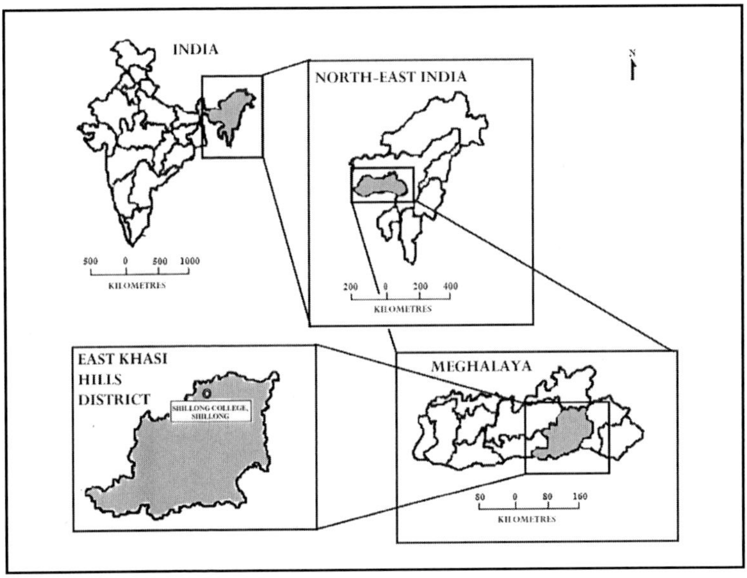

Fig. 6.1: Location Map of Study Area.

Questionnaire Survey

The information regarding students' knowledge of, perceptions and attitudes towards sacred groves was obtained using a structured questionnaire, which was comprised open and closed ended questions. A previous questionnaire devised by Diengdoh (2022) (Diengdoh and Jayakumar, 2021), was drawn upon, modified and adapted to frame a questionnaire designed to acquire: (i) the basic socio-economic and academic data of students, (ii) their knowledge and awareness of sacred groves, and (iii) their beliefs and attitudes towards sacred groves. Whereas the questionnaire tool of the previous study (Diengdoh and Jayakumar, 2021; Diengdoh, 2022) was designed to survey a sample population of villagers living in close association with a sacred grove, the present questionnaire has been designed for a population of undergraduate and post-graduate students enrolled in an institution of higher education. Therefore, the questionnaire was designed accordingly. The approach of Ormsby (2013) was also drawn upon during the process of questionnaire design. Students' attitudes were evaluated *via* questions and *via* their responses to various statements provided in the survey form to which they indicated their concurrence or opposition *via* their selection of fixed responses along an ordinal scale known as the Likert scale (Likert, 1932). The Likert scale has been used in wide variety of contexts, and is mainly implemented to measure human perceptions and attitudes towards a wide assortment of schools of thought, theories, ideas, physical objects among other things (Balasubramanian, 2012). In the present study, a 5-point Likert scale was employed, wherein students had to indicate whether or not they agreed with a statement and to what extent *via* the selection of anyone of the following responses (for each statement: "Strongly Agree, Agree, Neutral, Disagree, Strongly Disagree) which they felt was most representative of their views with regards to said statement (Willits *et al.*, 2016; Taherdoost,

2019). Preston and Colman (2000) indicate that 2–3-point scales returned lower (though still reliable) coefficients for Cronbach's Alpha reliability (Cronbach, 1951), and thus, to optimize the reliability of the results, a 5-point Likert scale was employed. The implementation of a 5-point Likert scale was also found to be an effective method of ascertaining respondent attitudes and perceptions regarding sacred groves (Diengdoh and Jayakumar, 2021; Diengdoh, 2022).

The questionnaire was reformatted as a Google® form (URL - https://forms.gle/VWAgidwxgn7w1zup9) and distributed to students of various streams, and in various stages of their course of study using WhatsApp® during the month of June 2024. Cronbach's Alpha reliability test (Cronbach, 1951; Preston and Colman, 2000) was used to assess the reliability of the data obtained *via* the use of the 5-point Likert scale.

Results

Socio-Economic and Academic Characteristics of Respondents

A total of 484 students comprising approximately 12.8 per cent of the total student population (inclusive of undergraduate and post-graduate students) of the college were sampled. Students came from either rural or urban areas. The origins (urban or rural) of students along with their gender distribution are shown in the Figure 6.2.

In terms of ethnicity, 95.2 per cent identified as tribals, while the remaining 4.8 per cent stated they were non-tribal. A large majority of students (88%) indicated that they are from Meghalaya, 5 per cent stated that they are from Manipur, while the remainder (<2% each) are from Arunachal Pradesh, Nagaland, Mizoram, Assam and Bihar, while 0.4 per cent did not specify their State of Domicile. The distribution of respondents

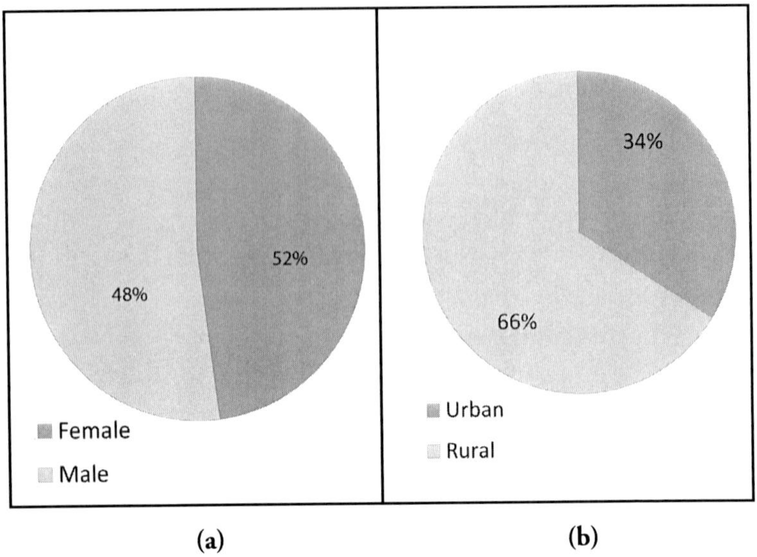

Fig. 6.2: Distribution of respondents according to (a) Gender and (b) Origin.

according to their religion/worldview is shown in Table 6.1.

Table 6.1: Distribution of Respondents by Religion/Worldview

Religion/Worldview	Percentage
Hinduism	5.4
Christianity	80.6
Islam	0.2
Buddhism	2.3
Niam Khasi	7
Niamtre Jaintia	2.5
Others	1.4
Prefer not to say	0.6

Respondents were also enrolled in a variety streams, semesters of study though a large majority were enrolled in undergraduate courses; they also hailed from a wide range of economic backgrounds, as is shown in the Figure 6.3.

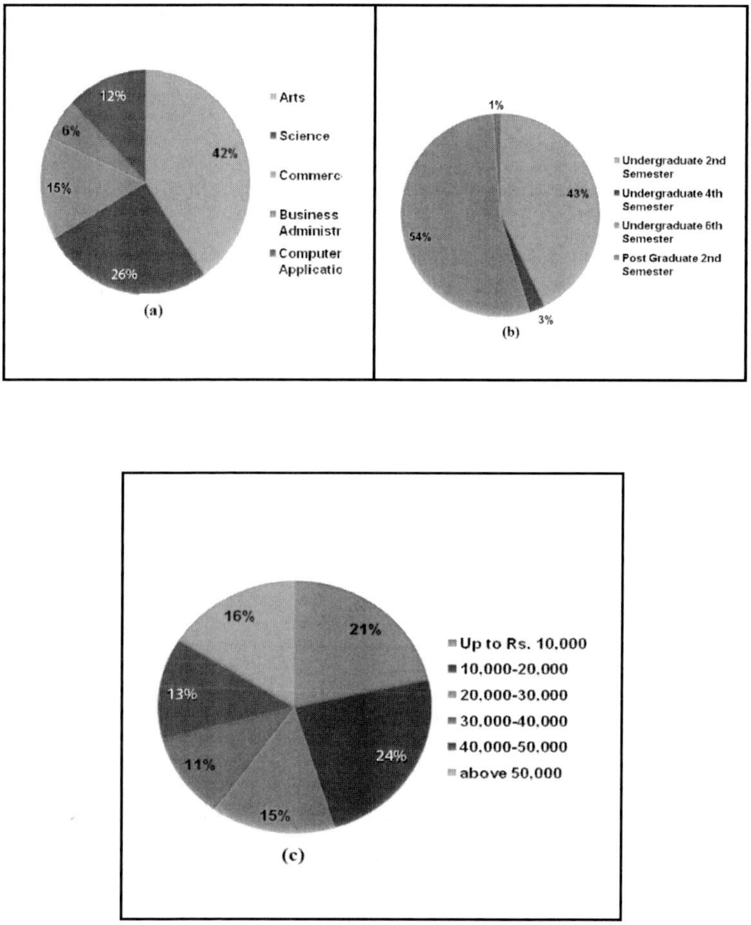

Fig. 6.3: Distribution of Respondents according to (a) Stream, (b) Semester of Study, and (c) Family Monthly Income Bracket.

Student Knowledge and Awareness of Sacred Groves

The students' knowledge and awareness of sacred groves as evaluated through several closed ended questions. The notable majority indicated awareness with the term *'sacred grove'* and a slightly greater majority stated they have knowledge of what a sacred grove is, although only a small minority stated they had visited the sacred groves. Their responses are shown in that Table 6.2.

Table 6.2: Students' Responses to Questions regarding their knowledge and awareness of sacred groves

Question	Yes (%)	No (%)	Not Sure (%)
Are you familiar with the term 'sacred grove'?	79.5	10.3	10.1
Do you know what a sacred grove is?	80.2	9.5	10.3
Have you ever been to a sacred grove?	53.5	43	3.5
Are there any sacred groves near the place where you live?	43.4	34.8	21.7
Are you aware of any rules or regulations put into place regarding the use of sacred groves?	65.5	14.9	19.6
Are sacred groves important in your opinion?	84.7	2.5	12.8
Do you believe sacred groves are declining?	43.8	14	42.1

Respondents were asked what, according to them, sacred groves are believed to be, and given the option to select multiple answers. Respondents went on to select multiple responses. Their responses are recorded in Figure 6.4.

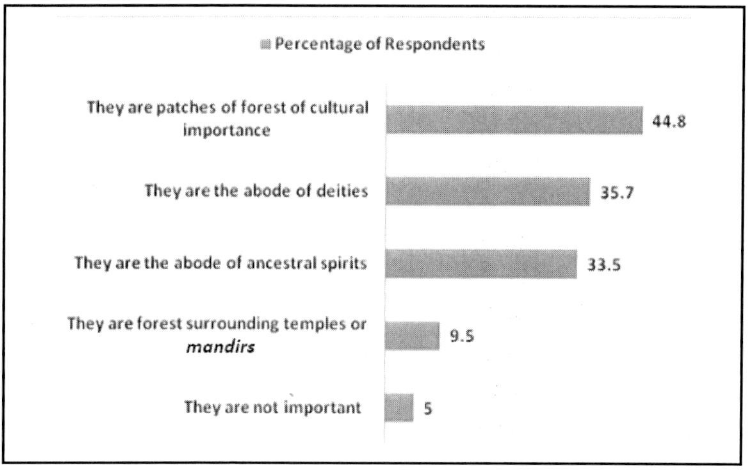

Fig. 6.4: Respondents' views on what Sacred Groves are believed to be.

Students who stated that they visit the grove were asked to give reasons for their visit. Their responses are shown in the Figure 6.5.

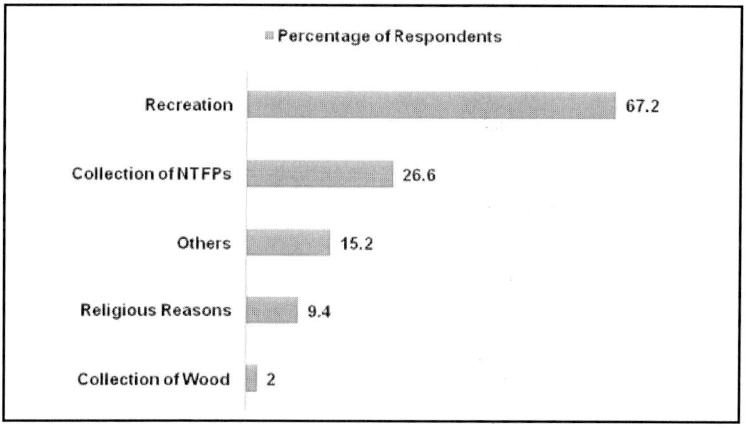

Fig. 6.5: Reasons for visiting Sacred Groves according to Respondents (%).

When asked to name any sacred groves near their residence, over 30 per cent of respondents named Mawphlang sacred grove, at Mawphlang, East Khasi Hills District, Meghalaya, including those who reside in Shillong, which is approximately 25 km away from Shillong. Other groves named include *Ïalong* sacred grove, *Law Lyngdoh* (*Sohra*), *Mawsynram* sacred grove, *Peinlang*, *Mynsoo* sacred grove, *Lum Symper*, Nongwar village, Pynursla, *Khlooblai, Khlookopati*, located in Meghalaya, *Koubru* in Manipur, *Kamalanagar* in Mizoram, and one gave the general location of Rajasthan.

Regarding awareness of the existence of rules and/or regulations instituted to preserve sacred groves and govern their usage, a majority indicated that they are aware of the existence of such rules (Table 6.2). Respondents were asked what taboo violations of sacred groves they have knowledge of and given the option to select multiple types. When further asked if they were aware of any violations of said rules, 49.2 per cent indicated that they are indeed aware of such violations while 32 per cent stated they are not sure, 15.7 per cent stated they are not, and 3.1 per cent denied the existence of such rules or violations. Those who indicated their awareness of such violations were asked to select the types known to them. Their responses are shown in Figure 6.6.

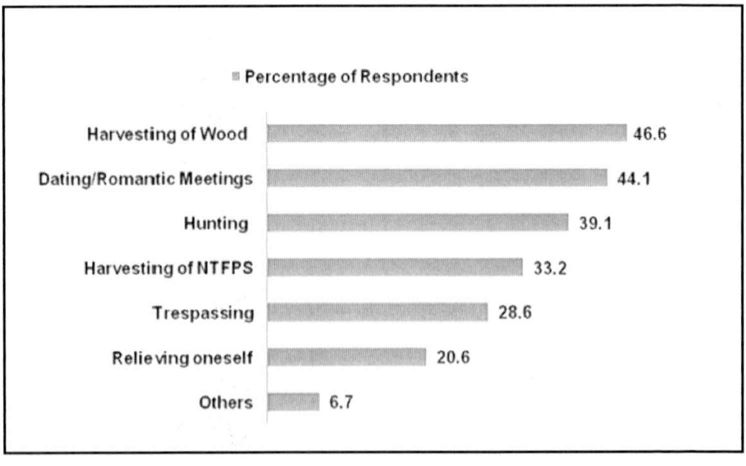

Fig. 6.6: Types of Taboo Violations according to Respondents.

When asked regarding their awareness of punishments given to any violators or perpetrators of unacceptable acts within sacred groves, the responses given by respondents were varied, and are shown in the Figure 6.7.

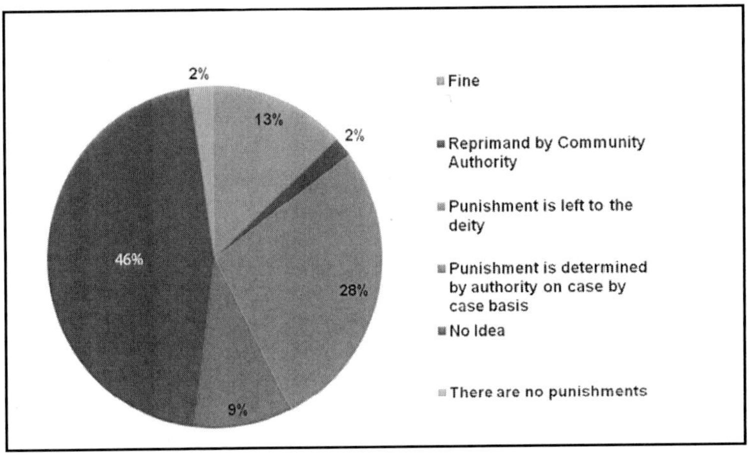

Fig. 6.7: Punishments given to violators according to Respondents.

Respondents who indicated that they consider sacred groves to be of importance were asked to give reasons for their answer. A number of them selected multiple reasons. Their responses are shown in Fig. 6.8 (a). Respondents who stated they believe sacred groves to be in decline, or were asked to give reasons for their answer. More than one reason was selected by many respondents. The reasons given are shown in Figure 6.8 (b).

Students who were unsure of whether or not sacred groves are in decline nevertheless gave reasons for the possible decline of sacred groves. Of these, the maximum number of students (56.4%) could not give a reason, 17.2 per cent stated there are other reasons, 9.8 per cent gave change in religious belief as a reason, 7.8 per cent listed climate change, another 7.8 per cent listed loss of interest among communities, 7.4 per cent expressed their belief that change in cultural norms are the cause, 2.5 per cent stated that increase in area under housing is a cause while only 2 per cent gave shifting agriculture as a reason. Only 1 per cent gave increase in agriculture as a cause, while even less (0.5%) stated that resource pressure is a cause.

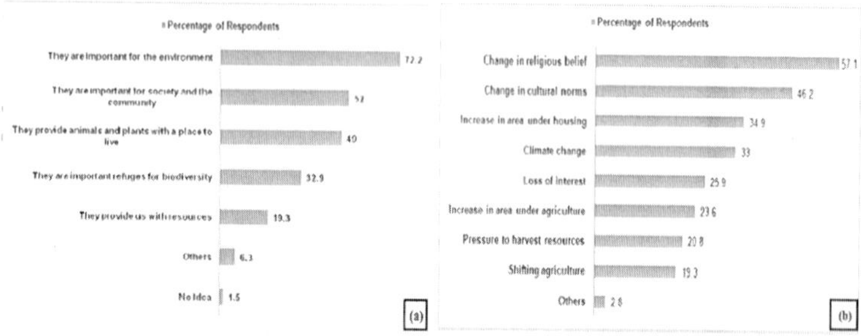

Fig. 6.8: (a) Reasons for the Importance of Sacred Groves (%) (b) Reasons given for the Decline (%).

Responses to Likert Scale Statements

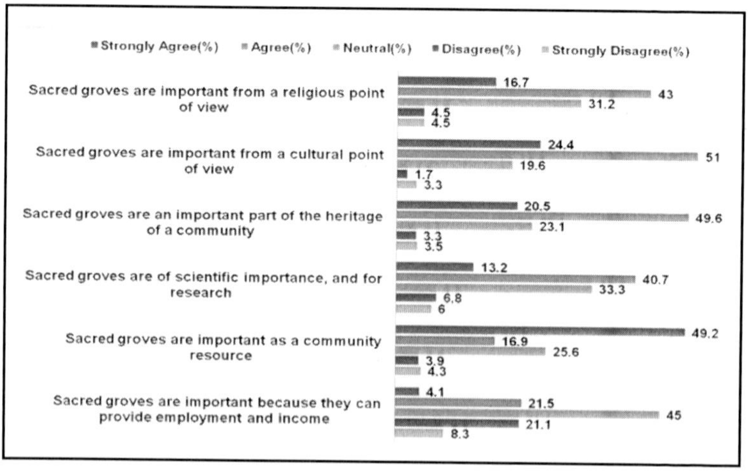

Fig. 6.9: Respondents' Attitudes and Perceptions of Sacred Groves.

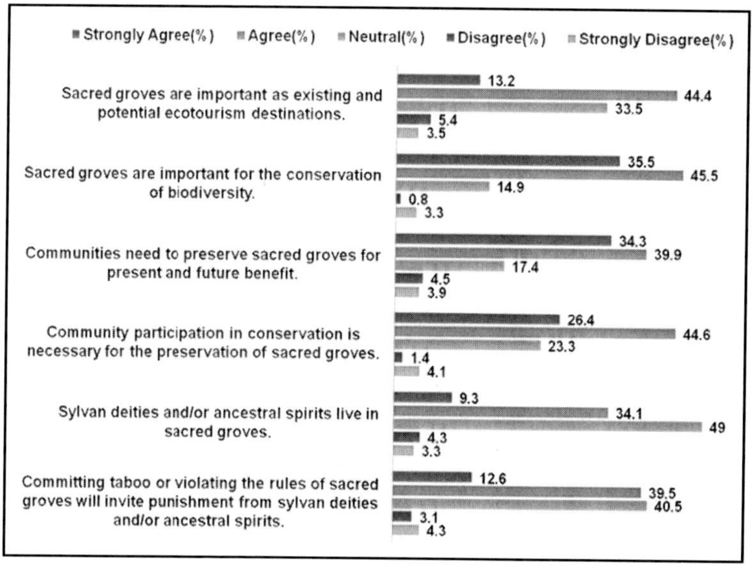

Fig. 6.10: Respondents' Attitudes and Perceptions of Sacred Groves Continued.

Respondents showed notable variation in their responses to various statements set alone a 5-point Likert Scale (Likert, 1932). Cronbach's Alpha Reliability Test (Cronbach, 1951) for 5-point Likert Scale data returned a value of 0.891, which, being higher than 0.7 indicates that the data is highly reliable. Respondents' attitudes as recorded by their responses to statements on a 5-point Likert Scale are shown in the Figures 6.9 and 6.10.

Discussion

Sacred Grove and the Role of Religion

Although the large majority (80.6%) of respondents identified as subscribing to Christianity, it is notable that a majority of them (57.5%) agree that sacred groves are of religious importance. As stated above, sacred groves in the form of church forests have been recorded in Ethiopia (Cardelús *et al.*, 2013; 2017; 2019; Abbott, 2019). In Italy, a Monteluco di Spoleto grove, held to be sacred during the Roman period has since come to be similarly regarded as such now in association with a nearby convent (Frascaroli, 2013). Furthermore, though only 42.3 per cent of Christians indicated agreement with the statement that sacred groves are the dwelling places of sylvan deities and/or ancestral spirits, it is remarkable that 52.1 per cent of them agree that said deities and/or spirits can mete out punishment to anyone who violates the taboos of a grove. Christianity, though a monotheistic faith, founded upon belief in the Triune nature, sovereignty and sole divinity of the god Yahweh, does not preclude belief in the existence of other spirit beings, which, though not considered divine or ontologically or economically equal to Yahweh, nevertheless may exert influence in various contexts to differing extents under His overarching authority. However, the discussion of the relationship between Christian monotheism and belief in other spirit beings falls beyond the scope of this chapter and further elaboration on the topic has been undertaken by other workers (Heiser, 2015; 2018; 2020). Previous work has recorded belief in the existence and influence of sylvan deities or spirits among Christian community members (Diengdoh and Jayakumar, 2021; Diengdoh, 2022), even though many communities are of the opinion that a change in belief system is a major cause for the decline of sacred groves, and by extension, in their conservation (see Tiwari, Barik and Tripathi, 1998; Ormsby, 2013), a finding supported by the present study (Fig. 6.8b). A large majority (76.5%) of respondents belonging to the indigenous religion of the Khasi tribe, the *Niam Khasi* indicated agreement that sacred groves are important from a religious point of view. However, a smaller number (64.7%) believed that sacred groves are the abodes of deities and/or spirits, but it is, however, remarkable that a greater number (70.6%) of them share the belief that taboo violations would incur the wrath of the said deities and/or spirits. A similar majority (75%) of those belonging to the indigenous religion of the Jaintia tribe, the *Niam tre* also held the opinion that sacred groves are of religious importance, though only 33.3 per cent hold to the belief that sacred groves house deities and/or spirits, while 50 per cent believe in that those who violate taboos will be punished by said deities and/or spirits. It is also remarkable that only 35.7 per cent and 33.5 per cent of respondents stated that they believe sacred groves to be abode of sylvan deities and of ancestral spirits respectively (Fig. 6.4), although 43.4 per cent indicated agreement with the Likert statement to the same effect (Fig. 6.9). This disparity may be due to a variety of reasons, such as respondents' believing that said deities may exist but not solely reside in said grove, or due to a form of psycho-spiritual compartmentalization, or cognitive dissonance whereby incompatible, unreconciled or contradictory beliefs may be held by the same individual or may be due to respondent error in responding to the statement. Various studies conducted in a number of areas have found that punishments doled out by spirits, according to respondents can include forms of body contortion in Meghalaya (Orsmby, 2013), to a variety of illnesses (Ormsby, 2013; Allendorf, 2014), crop damage (Acharya and Ormsby, 2017), or even death, either for the perpetrator or for a relative of theirs (Malhotra *et al.*, 2001). Among Hindus, 46.2 per cent agreed that sacred groves are the dwelling places of deities and/or spirits, though 50 per cent remained neutral and only 3.8 per cent indicated their disagreement. Sacred groves are known to exist in Hinduism as well (see Chandrashekhara, 2011a, 2011b; Ormsby, 2013), and the existence of spirits or deities falls well within the pantheistic paradigm within which Hinduism may be classified. However, only 34.6 per cent indicated a belief that acts of violation of taboo would invite punishment from the deity or spirit of the grove. The majority (63.6%) of Buddhists indicated neutrality to the statement that sacred groves house spirits or deities although fewer (45.5%) held the same view with regard to deities/spirits punishing violators. Only one identified as being a follower of Islam (0.2%) and this individual indicated a complete rejection of the above statements. In a study conducted in Yunnan province in China, the majority (94%) of respondents indicated that sacred forests benefit their community from a religious perspective, with most regarding their groves as places to engage in ritualistic burning of incense and in prayer for good fortune (Allendorf, Brandt and Yang, 2014). However, the present results must be interpreted with caution due the sample size for religions/belief systems other than Christianity being notably much lower by comparison.

The Role of Origin, Ethnicity and Gender

The majority of both rural (77.4%) and urban (83.7%) respondents stated they were familiar with the term sacred groves, and notably, 78.9 per cent of the former and 82.5 per cent of the latter stated that they know what a sacred grove is. This slight discrepancy may be due to respondent error during the process of filling up the survey form though it is conceivable that some respondents may be familiar with the term but may not unsure of what a sacred grove is. Again, similar percentages of rural (51.3%) and urban (57.8%) respondents stated that they had been to a sacred grove, while a lower percentage of rural (44.3%) and urban (41.6%) respondents indicated that there are sacred groves in close proximity to the place where they live. As previously stated, however, over 30 per cent of respondents living in Meghalaya gave Mawphlang sacred grove including many residing in Shillong, approximately 25 km away from the grove.

The majority of respondents indicated that they are of tribal ethnicity. Of them, 81.1 per cent were aware of the term 'sacred grove', 81.6 per cent stated they know what a sacred grove is. By contrast only 47.8 per cent of non-tribals indicated their awareness of the term, while 52.2 per cent stated they know what it is indicating they may be aware of the concept by other names, most likely in their respective local vernacular. It is also possible that this may be due to respondent error in answering the questions in the form. Notably, 85.7 per cent of tribal respondents indicated that they think sacred groves are important, though only 53.4 per cent indicated that they think they are in decline. By contrast, a smaller majority (65.2%) of non-tribal respondents think they are important and 52.2 per cent think they are in decline. Notably, 41.9 per cent of tribals and 47.8 per cent of non-tribals stated they are not sure if sacred groves are in decline. However, due to the fact that the overwhelming majority (95.2%) of respondents are of tribal origin, these results should be viewed with caution.

In the present study, student responses showed little variation with gender. A majority of male (85.7%) and female (73.9%) indicated they are aware of the term 'sacred grove' and a similar majority (82.6% of males and 77.9% of females) claim to know what a sacred grove is. A notable majority (87.8%) of males and females (81.8%) stated they consider sacred groves to be of great importance. Though, a greater percentage of males responded in the affirmative as compared to females as described above, overall gender was not found to be a significant factor in determining student responses.

Sacred Groves as Sources of Income and Tourist Destinations

A notable percentage of respondents listed Mawphlang sacred grove, a well known tourist destination as an example of a grove near them, however, 57.6 per cent of respondents indicated their agreement with the statement that sacred groves are important as sources of income, employment and as tourist destinations, a notably lower percentage than was found by Diengdoh and Jayakumar (2021), during whose study, 91.7 per cent were found to be in agreement, 8.3 per cent were neutral and 0 per cent indicated disagreement at Mawphlang (Diengdoh, 2022). However, they recorded 66.9 per cent of respondents at another village, Swer, within the boundaries of which another grove, *law Ryngkew lum Swer* is located as agreeing with the statement (Diengdoh and Jayakumar, 2021; Diengdoh, 2022). Furthermore, in the present study, only 25.6 per cent of respondents agreed that sacred groves can provide income and employment, while 45 per cent were neutral and 29.4 per cent indicated their disagreement. This finding contrasts the findings of the previous study at Mawphlang in which a majority of respondents (83.5%) at Mawphlang agreed that sacred groves provide employment but it is in keeping with the findings at Swer wherein only 10.1 per cent indicated they agreed with the statement (Diengdoh and Jayakumar, 2021; Diengdoh, 2022). It is to be noted that some residents at Mawphlang are engaged in the management and preservation of the grove as tour guides, entry counter staff etc. such that the status of the grove as a tourist destination provides them with a source of income, while at Swer, the grove is not a tourist destination and does not generate income (Diengdoh and Jayakumar, 2021; Diengdoh, 2022). In the present study, however, it must be noted that the selected sample population is comprised the students of an educational institution, not a village that administers a sacred grove. As such, the students sampled came from a variety of geographical locations, socio-economic, religious, cultural and ethnic backgrounds. The survey also did not focus on a population of whom a known number were engaged in sacred grove related tourist activity. It is, therefore, highly probable that these factors have played a role in the lower percentage of respondents agreeing that sacred groves are important as sources of income, employment and as tourist destinations.

Views on the Importance of Sacred Groves, their Decline and the Need for Conservation

The present study has found that an overwhelming majority (84.7%) of respondents think that sacred groves are important, a finding which is in keeping with previous work (Diengdoh and Jayakumar, 2021; Diengdoh 2022). However, though this previous study conducted among villagers of Mawphlang and Swer villages found that only a minority of respondents (20.3% at Mawphlang and 1.4% at Swer) think that sacred groves are in decline (Diengdoh & Jayakumar, 2021; Diengdoh, 2022), 43.8 per cent of respondents in the present study believe they are in decline while 42.1 per cent are not sure while only 14 per cent believe they are not. Previous studies have elucidated the impact of human disturbance on sacred groves (Tiwari, Barik, and Tripathi, 1998; Mishra *et al.*, 2004; Diengdoh, 2022), and in Meghalaya, it has been found that only a minority (less than 15%) of groves in the state can be classified as undisturbed (Tiwari, Barik, and Tripathi, 1998). The change in belief system has been regarded as a major contributing factor to the decline of sacred groves (Tomalin, 2004; Tiwari, Barik, and Tripathi, 1998; Sarfo-Mensah *et al.*, 2010; Ormsby, 2013; Diengdoh, and Jayakumar, 2021; Diengdoh, 2022), with some workers reporting decline occurring only with the onset of religio-cultural change (Wadley and Colfer, 2004), though some results may be described as mixed and must be interested cautiously (Byers, Cunliffe and Hudak, 2001). In light of this, it is highly probable that sample populations hailing from particular villages under whose jurisdiction particular groves fall, such as those at Mawphlang and Swer (Diengdoh and

Jayakumar, 2021; Diengdoh, 2022) have given their opinions on the decline of sacred groves only in relation or with reference to the groves that fall within said jurisdiction. In other words, it may be that the majority of respondents did not regard their own respective groves as in decline and thus answered the corresponding question based on this view. However, in the present study respondents, being from a wide variety of areas, some in close proximity to sacred grove and others not so, have in all probability given their opinion based on their general understanding of the decline of sacred groves.

Opinions were divided on the view—whether sacred or not sacred groves are in decline (Table 6.2), a finding which shows notable resonance with findings in Yunnan, southern Tibet, China as well, wherein a perception was found among just less that 50 per cent of respondents that sacred groves are in fact increasing in terms for growth (Allendorf, Brandt and Yang, 2014). In the present study, among respondents who are of the belief that sacred groves are declining, the majority gave change in religious belief as a reason, followed by change in cultural norms, increase in areas under housing, climate change, loss of interest among communities, loss of area under agriculture, resource pressure, shifting agriculture and those who selected others (Fig. 6.7b). These findings show notable agreement with those of Tiwari, Barik and Tripathi (1998) and Ormsby (2013) that indicate the interplay of multiple factors, chiefly change in religious beliefs, in culture along with a loss of interest among communities, notably among members of the younger generation (Tiwari, Barik and Tripathi, 1998), a significant deviation from those of Diengdoh and Jayakumar (2021) (Diengdoh, 2022) who found the most cited reasons to be climate change (33.3%), resource pressure (33.3%) and construction activities (33.3%) at Mawphlang. At Swer, however, only two respondents stated that they regarded sacred groves to be in decline, citing tree felling for the collection of wood as a reason (Diengdoh and Jayakumar, 2021; Diengdoh, 2022). It must be noted, however, that in the present study, the sample population is comprised exclusively of students enrolled at an institution of higher education, while in the former, the sample population were comprised of community members from all backgrounds and age groups residing in the villages located in close proximity to selected sacred groves.

The most cited reasons given by respondents for the importance of sacred groves was environmental, followed by social and community importance, habitat for wildlife and as refuges for biodiversity, and as sources for resources (Fig. 6.7a) Findings at Mawphlang showed similarity in that the most cited reason (37.3%) was that sacred groves perform ecosystem services and provide NTFPs, the former of which may be interpreted as being of environmental importance (Diengdoh and Jayakumar, 2021; Diengdoh, 2022). Their importance for conservation and as habitats for wildlife constituted the most cited reason (43.1%) at Swer (Diengdoh and Jayakumar, 2021; Diengdoh, 2022). Social and cultural importance were cited by 33.9 per cent of respondents at Mawphlang and 40 per cent of respondents at Swer (Diengdoh and Jayakumar, 2021; Diengdoh, 2022). Student perceptions regarding the importance of sacred groves in the present study may thus be described as being multifaceted, a finding which is in keeping with the findings of some previous studies conducted among a variety of populations (Allendorf *et al.*, 2006; Allendorf and Yang, 2013; Diengdoh and Jayakumar, 2021; Diengdoh, 2022) but which contradicts the findings of Allendorf, Brandt and Yang (2014) which describe their respondents' views of sacred groves as being *'one-dimensional'* (Allendorf, Brandt, and Yang, 2014).

The findings of the present study indicate that young individuals largely demonstrate an awareness regarding sacred groves and consider them to be of importance, a finding which contradicts those of earlier studies which found that beliefs regarding sacred groves have been largely relegated to the realm of superstition (Tiwari, Barik and Tripathi, 1998), or that the youth do not respect the age-old beliefs associated with them (Byers, Cunliffe and Hudak. 2001). While loss of interest was given as a reason for the decline of sacred groves (Fig. 6.7b), it was not the most cited reason. Nevertheless, while students may hold an opinion that sacred groves are important to them, the expression of this view in the form of practices that promote the conservation of the same are beyond the scope of study and assessment of the present study.

A majority (74.2%) of respondents agreed that sacred groves should be conserved so that present and future generations may benefit from them, a number which, while not as high as those recorded at Mawphlang (95.5%) and at Swer (97.1%) (Diengdoh and Jayakumar, 2021; Diengdoh, 2022), nevertheless indicates that a majority of students are of the opinion that sacred groves are of importance and must be conserved. As stated above, the intimate association between sample populations and their corresponding sacred grove in all likelihood, a contributing factor to the results found at Mawphlang and Swer (Diengdoh and Jayakumar, 2021; Diengdoh, 2022), in comparison to which, the results of the present study conducted among a random population of students of an educational institution, for whom association with a sacred grove was not a deciding factor in the selection process, may be regarded as notable.

Conclusion

Sacred groves occupy a distinct position in the socio-cultural and/or religious milieu of many communities, having arisen out of a blend of religious or spiritual belief, culture, tradition and community-based conservation. The youth of every community are regarded as the future custodians and caretakers of their socio-cultural inheritance, inclusive of sacred groves. It has been observed in past studies that socio-cultural and religious change manifesting in a community is also followed by corresponding changes in how sacred groves are perceived and managed (Tiwari, Barik and Tripathi, 1998; Sarfo-Mensah *et al.*, 2010; Ormsby, 2013). In the present study, however, an attempt has been made to ascertain the perceptions regarding sacred groves among college students, some of whom hail from communities in close proximity to sacred groves (43.4%), and others of whom do not (34.8%). While a large majority are aware of what sacred groves are (80.2%), and are familiar with the term (79.5%), a considerably smaller percentage (53.5%) indicated that they have visited one. Nevertheless, a majority of students (84.7%) indicate that they consider sacred groves to be of importance, while a similar majority (81%) agree that sacred groves are important for the conservation of biodiversity, and that there is a need for conserving them for the benefit of existing generations and for posterity as well (74.2%). Awareness regarding the socio-cultural, religious and biological importance of sacred groves among youth is important to the conservation of sacred groves through various avenues such as ecotourism, and the promotion of said sacred grove conservation ventures *via* various online and offline media. To this end, assessing the perceptions of students regarding sacred groves is important for ascertaining their views on sacred groves. It is the opinion of the author that studies in this regard are comparatively scant and that there is scope for the replication of the present study in other institutions and states in the region. The present study may lend impetus to further efforts to study the views, attitudes and perceptions among students regarding sacred groves in the region.

Acknowledgements

The author would like to thank the students of Shillong College for their cooperation during the data collection process. The author also feels indebted to Dr. B. Shabong, Dr. T. L. Nonglait and Dr. R. Kharkongor for their valuable input during the writing of this article.

REFERENCES

Abbott, Alison (2019). "Biodiversity Thrives in Ethiopia's Church Forests." *Nature, 565*(7741): 548-549.

Acharya, Amitangshu and Alison Ormsby (2017). "The Cultural Politics of Sacred Groves: A Case Study of Devithans in Sikkim, India." *Conservation and Society, 15*(2): 232-242.

Allendorf, T.D. and J. Yang (2013). "The Role of Ecosystem Services in Park—People Relationships: The Case of Gaoligongshan Nature Reserve in Southwest China." *Biological Conservation, 167*: 187-193.

Allendorf, Teri, Khaing Khaing Swe, Thida Oo, Y.E. Htut, Myint Aung, Keera Allendorf, Lee-Ann Hayek, Peter Leimgruber, and Chris Wemmer (2006). "Community Attitudes towards Three Protected Areas in Upper Myanmar (Burma)." *Environmental Conservation 33*(4): 344-352.

Allendorf, Teri, D., Jodi, S. Brandt, and Jian M. Yang (2104). "Local Perceptions of Tibetan Village Sacred Forests in Northwest Yunnan." *Biological Conservation 169*: 303-310.

Anderson, Danica M., Jan Salick, Robert K. Moseley, and Ou Xiaokun. (2005). "Conserving the Sacred Medicine Mountains: A Vegetation Analysis of Tibetan Sacred Sites in Northwest Yunnan." *Biodiversity and Conservation, 14*: 3065-3091.

Balasubramanian, Natesan (2012). "Likert Technique of Attitude Scale Construction in Nursing Research." *Asian Journal of Nursing Education and Research, 2*(2): II.

Byers, Bruce, A., Robert, N. Cunliffe, and Andrew, T. Hudak (2001). "Linking the Conservation of Culture and Nature: A Case Study of Sacred Forests in Zimbabwe." *Human Ecology, 29*: 187-218.

Cardelús, Catherine, L., Peter Scull, Joshua Hair, Maria Baimas-George, Margaret D. Lowman, and Alemaheyu Wassie Eshete (2013). "A Preliminary Assessment of Ethiopian Sacred Grove Status at the Landscape and Ecosystem Scales." *Diversity, 5*(2): 320-334.

Cardelús, Catherine L., Peter Scull, Alemayehu Wassie Eshete, Carrie L. Woods, Peter Klepeis, Eliza Kent, and Izabela Orlowska (2017). "Shadow Conservation and the Persistence of Sacred Church Forests in Northern Ethiopia." *Biotropica, 49*(5): 726-733.

Cardelús, Catherine L., Carrie L. Woods, Amare Bitew Mekonnen, Sonya Dexter, Peter Scull, and Berhanu Abraha Tsegay (2019). "Human Disturbance Impacts the Integrity of Sacred Church Forests, Ethiopia." *PLoS One, 14*(3): e0212430.

Chandrakanth, M.G. and Jeff Romm (1991). "Sacred Forests, Secular Forest Policies and People's Actions." *Nat. Resources J., 31*: 741.

Chandrashekhara, U.M. 2011. "Conservation and Management of Sacred Groves in Kerala." *Project funded by the Biodiversity Cell, Department of Forests and Wildlife, Government of Kerala.* KFRI, Peechi, Kerala.

Chandrashekhara, U.M. (2011). "Cultural and Conservation Values of Sacred Groves of Kerala, India." *International Journal of Ecology and Environmental Sciences, 37*(3): 143-155.

Cronbach, Lee J. (1951). "Coefficient Alpha and the Internal Structure of Tests." *Psychometrika, 16*(3): 297-334.

Dash, S.S. (2005). "Kabi Sacred Grove of North Sikkim." *Current Science 89*(3): 427-428.

Dash, S.S. and A.S. Chauhan (2002). "Kabi Sacred Grove in Sikkim. Relevance to Conservation." *Prospectives of Biodiversity,* Bishen Singh, Mahendra Pal Sing, Dehra Dun, pp. 713-731.

Decher, Jan. 1997. "Conservation, Small Mammals, and the Future of Sacred Groves in West Africa." *Biodiversity and Conservation, 6*: 1007-1026.

Deil, Ulrich, Heike Culmsee, and Mohamed Berriane (2005). "Sacred Groves in Morocco: A Society's 'Conservation of Nature for Spiritual Reasons'." *Silva Carelica, 49*: 185-201.

Diengdoh, Bennathaniel, H. (2022). *A Comparative Study of Biodiversity and Human Interactions in Two Sacred Groves in East Khasi Hills District, Meghalaya, India.* Ph.D. Thesis. Department of Ecology and Environmental Sciences, Pondicherry University, Puducherry.

Diengdoh, Bennathaniel Hynniewta and S. Jayakumar (2021). "A Comparative Study of Local Attitudes and Perceptions of Sacred Groves between Village Communities in East Khasi Hills, Meghalaya, India." *International Journal of Ecology and Environmental Sciences 47*(3): 197-208.

Dorm-Adzobu, Clement, Okyeame Ampadu-Agyei, and Peter G. Veit. (1991). *Religious Beliefs and Environmental Protection: The Malshegu Sacred Grove in Northern Ghana*. Acts Press.

Frascaroli, Fabrizio (2013). "Catholicism and Conservation: The Potential of Sacred Natural Sites for Biodiversity Management in Central Italy." *Human Ecology, 41*: 587-601.

Gadgil, Madhav and Vaman D. Vartak (1976). "The Sacred Groves of Western Ghats in India." *Economic Botany, 30*(2): 152-160.

Gupta, Bandana and Sanjay Sharma (2014). "Estimation of Biomass and Carbon Sequestration of Trees in Informally Protected Areas of Rajouri, J&K, India." *Int. Res. J. Environ. Sci, 3*(6): 56-61.

Heiser, Michael S. (2015). *The Unseen Realm: Recovering the Supernatural Worldview of the Bible*. Lexham Press.

Heiser, Michael S. (2018). *Angels: What the Bible Really Says About God's Heavenly Host*. Lexham Press.

Heiser, Michael S. (2020). *Demons: What the Bible Really Says about the Powers of Darkness*. Lexham Press.

Jyothi, K.M. and P. O. Nameer (2015). "Birds of sacred groves of northern Kerala, India." *Journal of Threatened Taxa, 7*(15): 8226-8236.

Khan, M.L., Ashalata Devi Khumbongmayum and R.S. Tripathi (2008). "The Sacred Groves and their Significance in Conserving Biodiversity: An Overview." *International Journal of Ecology and Environmental Sciences, 34*(3): 277-291.

Likert, Rensis (1932). "A Technique for the Measurement of Attitudes." *Archives of Psychology.*

Malhotra, Kailash C., Yogesh Gokhale, Sudipto Chatterjee and Sanjeev Srivastava (2001). "Cultural and Ecological Dimensions of Sacred Groves in India." *INSA,* New Delhi.

McWilliam, Andrew (2001). "Prospects for the Sacred Grove: Valuing Lulic Forests on Timor." *The Asia Pacific Journal of Anthropology 2*(2): 89-113.

Mishra, B.P., O.P. Tripathi, R.S. Tripathi and H.N. Pandey (2004). "Effects of Anthropogenic Disturbance on Plant Diversity and Community Structure of a Sacred Grove in Meghalaya, Northeast India." *Biodiversity and Conservation, 13*: 421-436.

Mohanta, R.K., A.K. Bhardwaj, B.S. Adhikari and P.K. Mathur (2009). "Distribution and Conservation Status of Sacred Groves (SGs) in Garo Hills, Meghalaya." *Indian Forester, 135*: 12.

Ormsby, Alison (2012). "Perceptions of Tourism at Sacred Groves in Ghana and India." *Recreation and Society in Africa, Asia and Latin America, 3*(1).

______(2013). "Analysis of Local Attitudes towards the Sacred Groves of Meghalaya and Karnataka, India." *Conservation and Society 11*(2): 187-197.

Ormsby, Alison A. and Shonil A. Bhagwat (2010). "Sacred Forests of India: A Strong Tradition of Community-based Natural Resource Management." *Environmental Conservation, 37*(3): 320-326.

Preston, Carolyn C. and Andrew M. Colman (2001). "Optimal Number of Response Categories in Rating Scales: Reliability, Validity, Discriminating Power, and Respondent Preferences." *Acta Psychologica 104*(1): 1-15.

Sarfo-Mensah, Paul, William Oduro, Ernestina Antoh Fredua, and Stephen Amisah (2010). "Traditional Representations of the Natural Environment and Biodiversity Conservation: Sacred Groves in Ghana". FEEM Working paper No. 87.

Taherdoost, Hamed (2019). "What is the Best Response Scale for Survey and Questionnaire Design; Review of Different Lengths of Rating Scale/Attitude Scale/Likert Scale." *International Journal of Academic Research in Management* (*IJARM*) 8.

Tiwari, B.K., Saroj K. Barik and R.S. Tripathi (1998). "Biodiversity Value, Status, and Strategies for Conservation of Sacred Groves of Meghalaya, India." *Ecosystem Health, 4*(1): 20-32.

Tomalin, Emma (2004). "Bio-divinity and Biodiversity: Perspectives on Religion and Environmental Conservation in India." *Numen, 51*(3): 265-295.

Tripathi, O.P. (2002). *Study of Distribution Pattern and Ecological Analysis of Major Forest Types of Meghalaya.* Doctoral Thesis. Department of Botany, North Eastern Hill University, Shillong, India.

Upadhaya, Krishna (2002). *Studies on Plant Diversity and Ecosystem Function in Sacred Groves of Meghalaya.* Doctoral Thesis. Department of Botany. North Eastern Hill University, Shillong, India.

Upadhaya, K., H.N. Pandey, P.S. Law and R.S. Tripathi (June 2005). "Plants of Ethnobotanical Importance in the Sacred Groves of Jaintia Hills of Meghalaya." *The Indian Forester*, Dehradun, *131*(6).

Wadley, Reed L. and Carol J. Pierce Colfer (2004). "Sacred Forest, Hunting, and Conservation in West Kalimantan, Indonesia." *Human Ecology*, *32*: 313-338.

Wessing, Robert (1999). "The Sacred Grove: Founders and the Owners of the Forest in West Java, Indonesia." *L'homme et la forêt tropicale*, pp. 59-74.

7

Khasi Traditional Institution to Mitigate the Impact of Climate Change

Barnes L. Mawrie

Introduction

Today the whole world is reeling under the negative impact of climate change. Climate change has become the most talked about topic in many international meetings and summits. We have seen a number of international conferences on climate change, the last being the COP29 which was held in Baku (Azerbaijan) in November 2024. However, many summits may take place but climate change remains an impasse until and unless there is a change of attitude, mentality and a conversion of the human heart. As long as humans do not respect nature, they will continue to exploit it. There needs to be a change of perspective towards our natural environment, to see it as "mother" and not as "matter", to understand our role as "care-takers" and not as "exploiters". Speaking about the Khasis' traditional respect for Mother Nature, it all boils down to their perception of nature and their consequent attitude towards her. For the Khasis nature is perceived as a "subject" to relate with and not as an "object" to be exploited.

Khasi Institution of Law Kyntang

Among the Khasis, nature is considered holy and sacred and for this reason, certain forests are marked as inviolable. There is an ancient practice among the people to declare certain groves as *Lawblei* or *lawkyntang* or as *Law Lyngdoh* or *Law Adong*. They all carry the same implications, namely that they are sacred and not to be violated by people in any manner. Some of the well-known sacred groves found in the Khasi and Jaintia Hills are *Law Lyngdoh Nongsangriang* in Nongstoin kingship, *Law kyntang Umwai* in Mawlong kingship, *Law Suidnoh* in Laitryngew, *Lum Ryngkew Swer* in Sohra kingship, *Law Lyngdoh Mawphlang, Law Lyngdoh Jowai, Lawblei Shyllong* and many others (S. Kharakor, 1989). We do not actually know the date of origin of such a significant and impactful practice. However, there are conceivable reasons behind such a traditional institution.

(a) *Theological reasons*: The Khasis since times immemorial have always displayed a great respect and love for their natural environment. They address nature as "*mei*" (mother) for they accept the fact that Mother Nature nurtures them and provides them with all the necessities of life from food and clothing to shelter and space. Like any tribal community, the Khasis have a religion that is elementally connected to their natural environment. In fact, the Khasi worldview can be said to be eco-theandric meaning to say that reality is trinitarian in nature, God-Man-Nature (Mawrie, 2009).

    The Khasis no doubt, have always had a strong eco-spirituality based on the concept of sacredness of nature.

Nature is seen as God's creation along with human beings. The existence of sacred mountains, hills, rivers and sacred groves or sacred plants vindicates this fact. For the Khasis, nature is a holy sanctuary where God resides and continues to interact with man. From this perspective, the ancient Khasis already practiced what Psalm 24:1-2 says, "the earth is the Lord's, and everything in it, the world, and all who live in it: for he founded it on the seas and established it on the waters." The Khasis treat nature with respect and they reveal a certain feeling of awe towards her. They believe that the vast nature surrounding them is the dwelling place of the spirits (good and evil), where the "*rngai*" (shadow) of the deceased roam about. Therefore, they seek constantly to appease such spirits so that they do not interfere with human life. For this reason, they offer appeasements and resort to divination in order to discern God's will in their life. This, however, should not be interpreted as spirit worship or idolatry, as some authors have mistakenly thought. The Khasis believe only in one God (*U Blei balah baïai, U Nongbuh u Nongthaw*) and Him only they worship. Nature on the other hand, is perceived as the dwelling place of God and thus it is sacred and ought to be revered.

(b) *Pragmatic reasons:* Socio-economically, the ancient Khasis realized that forests are sources for their survival from which they derive their food and drink, materials for shelters and even herbs for their medicines. This consideration is what led them to revere and value trees and plants. The ancient Khasis were a very wise and pragmatic community. They understood that left without control, the future generations would cut down trees wantonly. Therefore, to pre-emptively prevent such tragedy, they selected certain groves around their settlements and gave a religious sanction to them thereby calling them "*Law Kyntang*" or "*Law-blei.*" With such sanctions, it was impossible for the people to do anything that would desecrate these places without fear of divine retributions. Such pragmatism is what modern societies lack today. Instead, we have become victims of materialism and consumerism, which consequently lead to overexploitation of our natural environment (Mawrie, 2022).

Eco-significance of the Lawblei, Law Kyntang

The sacred groves among the Khasis carry a lot of socio-cultural, religious and ecological significance. It is believed that the deities live in these groves and, for this reason, no one is allowed to damage them by lumbering, mining, quarrying or by indulging in any other injurious activity. In these groves, the trees and plants grow without threat of being destroyed and the animals too enjoy the freedom and guarantee of security to life. Some of the sacred sanctions to be observed in these groves are:

(a) The visitor to these forests should not cause any damage whatsoever. He or she can only admire its beauty. Even plucking a flower, a leaf or cutting a branch is strictly forbidden. In short, the forests must be respected and everything must be left in their own proper places.

(b) Littering of these forests in different manners like throwing garbage or urinating is strictly forbidden. Such acts are considered as insults to the guardian spirits of these forests and so they can displease them.

(c) Anyone who visits these sacred grounds should observe decency in comportment. Use of foul language or cursing is considered a taboo in these places. Even impure and evil thoughts are not to be entertained while one is present in these places.

The person who dares to violate these sacred groves is brought to task by the village *durbar* for it is considered a crime (*ka sang*) not easily forgiven. It is also believed that violators of these sanctions will incur incurable diseases like insanity, loss of sight, lameness or some other psychological derangement, or even suffer unnatural deaths.

The sacred grove is a sort of divine sanctuary where the clan or a group of clans who take care of it, perform their periodic sacrifices to God the Creator. It is from this sacred place that they converse with their Creator through rituals and prayers, "*The sacred forests or Law kyntang can be called the churches or sanctuaries of the Khasis. Because it is here that man is able to enter into communion with God more than in any other places.*" The Khasis do not have temples or churches because they believe that the whole natural environment is God's dwelling place. The sacred grove (*lawblei-lawkyntang*) is considered by the Khasis as a paradise of the pristine glory of creation. It is here that nature is in a state of peace and harmony, free from any fear of destruction. Here men live in harmony with each other, with the animals and the birds, with the fish and the fowl, with the trees and the mountains and even with the spirits themselves.

Every sacred grove is guarded by *Ki Ryngkew Ki Basa*, the deities commissioned by the Almighty to reside in nature. The Khasis perform sacrifices of appeasement and other rituals to these deities as a sign of their gratitude to them who look after their villages and the natural wealth and protecting them from the evils beyond the knowledge and control of man.

The ecological significance of the *Law Kyntang* is something commendable. It shows the great respect and sense of responsibility that the Khasis have towards their natural environment. They truly show themselves to be the stewards and not the masters of creation, the ones who are called to uphold creation and not to destroy it. Fortunately, the Khasis today still maintain these sacred places in different parts of the region.

Relevance of Law Kyntang for Today

As mentioned earlier, we can never fight against climate change unless we change our mentality and attitude towards our natural environment. Definitely, the institution of the *Law Kyntang* is contributing significantly to mitigating the impact of climate change. However, the concept of *Law Kyntang* emerges from the people's perception of nature as sacred. This institution will remain and can even be propagated if people once again perceive the sacredness of mother nature and act accordingly towards her. In fact, we may say that the concept and practice of sacred groves is so relevant and necessary today more than ever. More communities in the world should nurture such attitude towards their natural environment and learn these basic lessons from our tribal communities (Mawrie, 2022). In order to preserve this valuable ecological institution, the following things need to be done:

(a) Conscientize the young generation about the sacredness of our natural environment. Perhaps the Church can do this task better than the secular institutions. Sanctions given by religions are more powerful than any government laws. There is a looming danger today that our tribal people are gradually losing the sense of the sacredness of Mother Nature. This is having a catastrophic impact on our natural environment. It is important that schools and colleges should also conscientize students regarding this.

(b) Both secular and religious institutions should encourage the preservation, maintenance and propagation of sacred groves wherever possible. Afforestation projects are to be encouraged and perhaps more *Law Kyntang* could be created all over our state and elsewhere as well. The governments should help financially in the maintenance and improvement of the existing sacred groves.

Conclusion

In conclusion, one can say that all of us are anxious about the adverse impact of climate change. We are witnessing the natural disasters that take place every year due to climate change. The question we may ask at this juncture is "what can we do about it?" Yes, we can do much about it. As tribal communities we are contributing significantly towards mitigating the impact of climate change through the institution of *Law Kyntang*. The Earth is an organism and therefore, whatever we do even the little that we can, will definitely have a positive impact on the planet. Bhutan is said to be a country with minus CO_2 and with surplus oxygen. This is because 70 per cent of the country is still covered with forests (Mawrie, 2022). Imagine if all our states in North East India were to maintain at least 60 per cent of forest land, what an ecological impact we would be making in the world. For this reason, the institution of sacred grove (*Law Kyntang*) must be preserved and encouraged in every community all over the world.

REFERENCES

Kharkongor, S. (1989). "Ki 'Lawblei' 'Law Kyntang", in *Seng Kut-snem: Nineteen Year Souvenir.* Shillong: Seng Khasi.

Mawrie, B. (2009). *The Khasis and their Natural Environment*. Shillong: VIP.

Mawrie, B. (2022). *Climate Change: A Theological-Moral Response.* Shillong: VIP & VIRTAS.

Mawrie, B. "The Sacred Forests and their Sustenance in the Khasi-Jaintia Hills". Unpublished.

8

Re-discovery of the Value of Indigenous Peoples' Ecological Resources

A Model for Environmental Preservation

S. Hayong

Introduction

Indigenous people are considered to be the first or original settlers of the land. As the original native settlers they have come to reside in the land before the entrance of any other people groups or other communities. They are the one who give names to the hills, the mountains, the rivers, the trees, the plants and other creation around them (Longchar, 2012). An act of giving names does not simply means to know them and differentiate them from other rather it convey and represent the value and importance of each and every species and resources. Moreover, giving names to every space and resources available portrays ownership while at the same time automatically provides a strong sense of attachment and close relationship to the nature and environment around them. In other words, it helps enhancing their knowledge about every pace, every pulse and rhythm of the nature. However, with the impact of external forces most of the indigenous wisdom and knowledge related to the nature around becomes blur and distorted and many have been diminished in their daily life and way of living. Hence, rediscovery of the value and importance of the indigenous people ecological resources is not only necessary but mandatory. These resources will offer important ecological model that could help preserving and protecting the nature and its surrounding from further ecological crisis.

Indigenous Peoples' Ecological Resources

Indigenous or native people have rich ecological resources that has been transmitted for many generations through the test of time. Some among many ecological resources that has been transmitted include land, forest and water. Indigenous people have a very close attachment and close relationship with the nature around them. Nature is not only their source of living, or their dwelling habitat but it is part and parcel of their life. Being part and parcel of their life, diverse elements of nature play important role in their entire way of living. In other words, nature provides them rich ecological resources that enhances their entire life reality and wide ranging experiences. Their long association with nature enables them to recognize the value of different natural and ecological resources like land, forest, and so on. They are in reality play important but sacred role to help all living and non-living things survive the challenges of time and space. They have a critical role concerning the life and survival of all that exist.

Ecological Resources and Mother Earth

For the indigenous people, mother earth is a home to all creation either living or non-living species. For them, mother earth offers a safe and secured habitat to each and every species that inhabit the planet. It is in the bosom of mother earth that all life forms exist and all things both living and non-living inherently obtain adequate space to dwell and reside in certain respective allotted space as well as to flourish and prosper in an orderly manner which need to be maintained and preserved. Indigenous people in diverse space and time possess certain valuable tales and stories relating to the role and functions offered by mother earth towards the flow of life in certain location. These stories highlight the abundant fundamental services delivered by the mother earth towards the growth and well-being of all lives in the planet. For instance, the creation story of the Khasis not only highlighted the origin of creation or the way each created things find their place in the realm of mother earth but at the same time it reveals how the mother earth play an active and dynamic role supporting the survival of all life forms in the planet irrespective of their size or importance (War, 1998: 41-43 and Nongkynrih, 2007:1-5). From the tiny plants to the big animals, all share one's own respective and particular space within the bosom of the mother earth. In this regard, Junita War rightly states when she mentioned "the loving and caring hands of mother nature reach out to all, great or small. The tiny ant has much the right to live its life as the huge elephant; the gentle deer can graze on the same meadow as the wild bison" (War, 1998). All created things both living as well as non-living could find their secured habitat in certain location within the sphere of mother earth. In simple term, they are not a part of but truly part and parcel of all the created whole having their respective rights and privileges as God's creation. Hence, it is accommodative, life affirming, nourishing, supporting and provides everything that is needed to sustain the growth of life and existence.

As it is a home to all creations, in reality it offers each species and life forms, essential abode to interrelate, interact and support each others survival encountering existential challenges down through the centuries. One may note that diverse life forms require each other supporting assistance necessary for their survival and subsistence amidst different circumstances and capacities. To state differently, each species possess certain assigned roles and purposes intended for the utmost good and well-being of all life forms within the inhabited earth. Moreover, the vibrant crucial functions and varied subtle facilities provided by each creation to the other has nonetheless help contributing towards the continuous flow of life cycle in the planet. It has in one way or the other equipping each life form to prosper and thrive in certain inhabited location. As it has mentioned elsewhere that "the unwanted waste of one being becomes the most wanted feed of another. The end of one process becomes the natural beginning of another" (Kozhamthadam, 2013). This cycle of life chains portrays the vital functions and dynamic confluence of everything that exist while at the same time could help assisting the entire flow of life series within the delicate space orderly allocated in diverse parts of the nature. It is not wrong to say that a wide varieties of all life forms or the so-called bio-diversities besides adorning beauty and richness to the natural surroundings they also provide "ecological harmony that makes life on earth a pleasant experience" (Kozhamthadam, 2013). Each species either big or small acquire certain vital necessary roles and functions in maintaining the ecological balance and harmonious peaceful co-existence of all the created things. In sum, to recognize this reality is nothing but to appreciate the intrinsic or core value and worth of each ecological resources.

The Value of Indigenous Peoples' Ecological Resources

As stated earlier, indigenous peoples around the world have rich ecological resources that has been preserved, maintained and transmitted down through the centuries. As they are the first settlers of the land they rather have better knowledge on the rhythm and flow of life in the entire natural surroundings. This illustrated the ground realities of their relational closeness and attachment with all natural entities sharing life realities in a broader community. It has in reality equip them to conceive the importance of maintaining and preserving natural surrounding not only for the need and requirement of human communities but for the entire broader community of all life forms. For them, each and every species or resources require certain space to abode, to grow and to prosper in its own way without any disturbance or intrusion by any disruptive forces that may disturb their very survival in their respective allocated habitat. Consequently, this realization could enable each life forms to provide and render better ecological necessary for the survival and the growth of life within the planet. To state differently, any attempt to disrupt and hamper this reality of life is not accepted and unwelcoming. For the indigenous people, their constant involvement with the reality of life in their long term association with nature since time immemorial not only help shaping their worldview but also provide them ample ways of understanding the importance and value of other resources. It is this kind of relationship that makes indigenous people to retain deep respect and value the importance of different earth resources. Hence, any effort to damage or spoil this relationship means self-destruction. In other words, to destroy the earth and its resources is also to destroy this relationship. It is like inviting trouble and misfortune not only to oneself but also to the entire humanity and non-human creations.

Indigenous People and Land

Traditionally, for the indigenous people, land plays an important role in their life. It is part and parcel of their life. One cannot separate land from the life and experience of the indigenous people. One may note that the moment they are separated it may perhaps left adverse and undesirable impact to both. This implies that while indigenous peoples depend on lands for diverse need and necessary essential requirement of daily existence, land needs someone to look after and preserve its overall maintenance and very survival confronting the emerging challenges and the test of time. The creation story of the Khasis reveals that when everything was created and situated in perfect and systematic order and all life forms exist and began to prosper, the mother earth or *ka Mei ramew* made a genuine request petition to God the creator or here in this case *U Blei Nongthaw Nongpynlong* to "send someone to be steward and controller of all living things" (War, 1998). It is due to this urgent plea of mother earth that God convene *Ka Dorbar Bneng* or divine assembly together with the sixteen families or *Ki KhadhynriewTrep Khadhynriew Skum* where they resolved to send the seven huts or the *Hynniew Trep Hynniew Skum* out of the sixteen families to come down to this earth and take care of the entire creation of God (War, 1998). Here, an act of appeal to God the creator for stewardship of the earth and its resources has two crucial importance and significance. First, it depicts the close relationship and attachment the mother earth has to its creator. Second, it reveals the inherent value and intrinsic worth that all ecological resources possess. These important values and worth is not inscribed or engraved by anyone but God the creator who made everything that exist. Indigenous people has well recognised and comprehended this reality attached to the mother earth and to the entire ecological resources. For them, all created things or ecological resources possess inherent right and value as intent by the creator from the beginning of time. It is such conception that enables them to realize and understand the value and worth of all created things.

Land as Mother

Indigenous people around the globe consider land as their mother. Generally, they call it mother earth. "Land as Mother" is a conceptual framework that encompasses in-depth and profound meaning penetrating every aspect of life reality. It is not wrong to say that there is no aspect of life that is not connected to land. To state differently, every aspect of life either social, economic, political, religious etc. are well connected to land (Mawrie, 1991 & 1997). It is in the bosom of mother earth that the entire life cycle and life process is enfolded. Accordingly, there is enough valuable space for each lifeform to begin, to grow and prosper amidst certain challenges that may perhaps come along the way. Being considered as mother, land not only provide them space to live but properly nurture them through diverse services provided by each and every life form. For the Khasi people, there are three important usage terms related to land as mother.

First, referring the earth as mother the Khasi people calls it as *Ka Mei Ramew*. The word *Ka Mei* means 'mother' and *Ramew* means "ground or terra firma". The ground is a place where they earn their source of living, a place where they cultivate their vegetables, fruits, medicinal plants, edible roots etc. It is a place where they received their house building materials, farming tools, and many other things needed for sustaining their life on this planet. Hence, land or ground is the most precious component that contains bounteous value and worth. Hence, they considered it as their mother. Moreover, *Ka Mei Ramew* is the creation of God, the creator or *U Blei Nongbuh Nongthaw* at the beginning of time. Being created by God, *Ka Mei Ramew* has its close attachment and connection with its creator. In short, whether humans exist or not, *Ka Mei Ramew* has its own value and worth in the sight of its creator. The creation story of the Khasi people depicts that the mother earth acts as co-creator with God by giving birth to other creations like the soil, animals, birds, plants, trees, grass etc. that are the source that sustain life (War, 1998). The concept of *Ka Mei Ramew* depicts the idea that mother earth has not only giving birth to different species but at the same time situate them in certain allocated space suitable for their life existence. There is adequate room for all creations either living or non-living irrespective of their size and varieties to live, to grow and prosper in particular location within the realm and domain of the mother earth. For instance, a small creature like insect has specific and actual space to live and the right to persist as the big creature like elephant.

Another term used by the Khasis for the mother earth is *Ka Meirilung Ka Meirisan.* This literally means a mother who nurtures and takes care of anything young and tender and looks after them even when they have grown up. Hence, it is not only giving birth to diverse creation but support their very existence in diverse ways and capacities. In short, nature as mother is one who nurtures, embraces, supports and fosters the growth of life here on earth. It has the power and capability to provide all essential life nourishment and important sustenance process to numerous life form either human or non-human species irrespective of their size, varieties, location etc. amidst difficult terrain and challenges confronted along the way. Different life forms receive valuable source of living from the mother earth in different capacities and location. Whether one realize it or not, the mother earth has in reality provide adequate valuable free services to everything needed for life survival of either human or non-human species.

Another concept used by the Khasi people for mother earth is *Ka Mei Mariang* (Hayong, 2020). The entire natural surrounding is understood as a space that embrace all creation both living and non-living. Being a mother, *Ka Mei Mariang* house all life forms both human and non-human providing them safe and secured habitat without any bias or discrimination. *Ka Mei Mariang* plays the role of a mother to look after each life form with a motherly care and protection. In reality it portrays that all life forms have their respective value and worth for the mother earth. Referring to this perception from the indigenous belief system one may agree that if each and every life is valuable for the mother earth it is rest assured that the same concern is applied to the marginalised poor and destitute within the community. To state differently *Ka Mei Mariang* will definitely taking necessary crucial care safeguarding the need of the marginalised section of the society. In other words, being a mother to all life forms, the mother earth has certain dynamic and significant role related to diverse need and requirement of the so-called underprivileged section of the community.

There is a term normally referred in Khasi, which says *Ki Khun ka Mariang* which literary means 'children of Mother Nature'. This phrase is referred to those children hailing from very poor background living in pathetic condition with no proper dwelling place or having no relative to look after them that has affected their overall life existence. This section of population are those members who depend mostly to the free services offered by the mother earth in different layers and aspect of life. They consist of those who do not have a proper dwelling house, having no proper food items to eat, malnourished, and even no proper cloth to wear. Most of the basic needs and necessity of daily living like proper healthcare facilities, important educational facilities and privileges of the day are beyond their reach. However, in a situation where desirable support and assistance seems to be a distant reality, it is *Ka Mei Mariang* or mother earth that is always there offering diverse important free services necessary to meet any sort of life challenges that could hinder their very survival. The term *Ki Khun ka Mariang* is also referred to the small and young children that rarely realise or realised their needs in certain situations. Nevertheless, despite their ignorance of their needs to protect themselves in particular circumstances, it is normally believed that the Mother Nature will take care shielding them from any problems and difficulties that may affect their life. For instance, if the children continue to play outdoor for a maximum period of time during the cold winter season, the Khasi used to say Mother Nature will provide them warm clothes.

The above concepts discussed explicates the reality related to the role and importance of mother earth not merely in providing space for each creation but also as the one who takes care, nurtures, fosters and brings up the entire creation to their full potentials. It illustrates the solemn duty of the earth as mother basically intent to endorse and promote the dynamic of life. Like a mother, the earth is always ready to deliver fundamental services that could help facilitate the better welfare of every life forms positioned in diverse location. From a small and tiny creature like the ant to the big and mighty creature like elephant all species are assigned certain specific sites suited for their survival. Hence, it is accommodative, life affirming, supportive, and empowering. It has all the compulsory details required for the sustenance and continuity of life. Having acknowledged the functional characteristic of mother earth in nurturing life through their close attachment with her in different capacities, indigenous people uphold the mother earth in high regards and reverence.

Land as Source of Livelihood

For the indigenous people the entire land and the surrounding environment not only offer them space for habitat but at the same time possess loads of diverse source for life sustenance. It is from land where all the essential needs and life resources necessary for daily living and life existence derive. To state differently, land offer them rich resourceful materials that could sustain life, comprising firewood, numerous wild fruits and vegetables, edible roots, medicinal plants, house building materials, and the like. In the distant past, prior to their contact with the outside world, plenty of life resources were easily available in nearby surrounding environment that perhaps be sufficient enough for their simple way of living. Even today most of these life sustenance resources are available in different market of the land either in urban or rural village market. A visit to any market place both urban and interior rural market may come across different varieties of seasonal wild fruits and vegetables generally display in particular allotted space meant for commercial purposes. Looking from the present day context where organic food materials not only scarcely available but in high demand in different capacities, these wild natural fruits and vegetables become the most preferred valuable food items earnestly desired by many today.

Nirmal Minz maintains that "land is life" (Minz, 1998). No land no life. Hence possession of a part of it means life. This statement is not merely to be understood in terms of commercial aspect of land rather it portrays land as the most precious commodities highly regards by many. For the indigenous people it speaks volumes of different layers of attachment and relationship they have with the land. Commenting on the importance of land to the growth and development of life K. Thanzauva rightly maintains when he says "as a baby is sucking the milk from the mother, human beings are sucking the milk of the earth in various ways" (Thanzauva, 2004). To state differently, besides being the sources of livelihood, land encompasses ample rich resources that could help sustaining life survival in different capacities. Hence, owning a portion of it adds value to someone's life and existence. It not only helps one own life prosper and flourished but also enables someone to endure the test of time.

Land as Source of Identity

For the indigenous people land not only provide them sufficient source of livelihood rather it is something that gives them identity and personhood. To have land or to possess a part of it means to have identity. There is huge difference between the landless individuals or the have nots and those who have land. For it is land that makes a difference between different sections of population. That is, to have land reflects a well established life whereas to have not may even comparable to refugees within one's own community. Commenting on the value and importance of land to the indigenous people, Nirmal Minz rightly argues when he states, "ownership of land and alienation from land has theological implications for the tribal" (Minz, 1998). Here, alienation from land for the indigenous people does not simply means being detached from the soil, but it also means detached from God its creator whom the indigenous people believes to abode in the land or *uba dap ha ki jakabaroh*. Some of the important points to be noted may be mentioned as follows:

First, there is a Khasi saying frequently referred which says *Ki la ia longbriew mynta* literally means, "at present they have become better human". This phrase is a common statement usually referred to those underprivileged individuals or poor families struggle to meet even the basic necessity of life, having no land or proper dwelling habitat but now they have a better and prosperous life. This is not simple a reference to a better life they are living now but it also illustrate having a better identity and personhood.

Second, as land gives them identity indigenous people mostly identify themselves by the place where they belong. Whereas other people from different communities or societies identify themselves by their names indigenous peoples prefer to identify themselves by the place or village they come from. J.N. Chowdhury (1996) maintains that prior to the coming of the British colonial rulers the Khasi people generally identify themselves by the name of their village as *U Nong-Mylliem* (residents of Mylliem village) or *U Nong-Sohra* and so on. Practically, it implies that it is always the land not a person's name that is more preferred by anyone to identify themselves. One may note that whenever two Khasi people meet for the first time they will introduce themselves not by their names rather by the place or village where they belong. Hence, the first question is not to ask the name of the concerned person but the place he or she come from. For instance, the first question will be *Phi dei naei*? Literally means where are you from? It depicts the importance of land and attachment to the land which they belong. In other words, their personal identity and how they identify themselves to other is closely attached to the place where they have been born as well as the landscape around they have brought up, which they consider their habitat.

Third, it may be noted that as land gives them identity it has a special value and worth in their life. Hence, selling of land or disposing any part of it is a non-issue in ancient time. It is reported that prior to the coming of the British colonial rule to these hills selling of land is unheard of among the Khasi people. Land is not for sale but it is to be protected and maintained. For instance, when the Britishers needed some portion of land for the establishment of their headquarters at Sohra as well as other necessary spaces related to the function of their government, the Sohra state did not sale land to them. Rather they provided them the land in exchange with certain portion of the British land located at the Sylhet area presently Bangladesh (Giri, 1998). This elucidates the fact that selling of land is not only unknown but not welcome by the indigenous people at that time. According to the indigenous belief system of the Khasi people selling of land especially the inherited land or *Ka Nongtymmen* that has been transmitted down through the ages is strictly prohibited if not a taboo. It is considered or equivalent to selling their very own identity. But in case there is a necessity urgently demand to do so, they will rather prefer to sell it within the clan itself. Because once the land is vended outside, the domain of the clan it will severely affect the clan as well as the future generations of the entire clan or the family (Hayong, 2018). It is not wrong to say that the Khasi people of the earlier times clearly understood the value and importance of land not only for the present day generation but for many generations to come.

Nature as Text: Indigenous Wisdom

Looking to the history and development of indigenous people one could acknowledge the role and importance of oral traditions that has pass on from generation to another. For many centuries back indigenous people around the globe do not have written books or records that place in written forms about their experience, culture, history and so on. However, most of the important teachings and customary practices are transmitted to the latter generation through oral tradition (Hayong, 2020). This has been preserved in the forms of stories, folklores, folktales, legends, myths, and so on. Nevertheless, they have so far generally occupied an important place and role in different aspects of life among many indigenous communities. Alongside in their journey nature have in reality provide them constant teaching and guidance enlightening the people in different situation and life experiences related to different realities of life. Nature possesses enormous source of knowledge, compiling immense data and distinct lessons that could help assisting the people with important direction, proper vision, purpose, moral guidance and goal to achieve.

For the indigenous people each and every creature either big or small could offer some important lessons for the people to be learnt on different situation and challenges confronted in life. From the tiny insects to the big and mammoth creatures has something good to offer to the people related to different circumstances they are facing along the way. Each creature can predict and foretell approaching or future events that may occur in the next few days. This will help the people to learn from these signs and get prepared themselves for diverse challenges ahead. For instance, when they see that the ants carry their food stuff from the lower to the upper places during summer it is understood that heavy rain will occur in the next few days. So also, when a particular bird cries it is a sign that something unfortunate like accidents and death may take place in the village. War further elaborate on this aspect when she says that the appearance of *ka niangkariang* or water beetle on the stones and trees along the riverside is a sign that foretell the coming of rain or not the coming days. So also, the emerging of *Ki Kber* or winged white ants from the ground indicates that rain will stop for a while (War, 1998). This implies that nature and the surrounding environment have certain lessons to educate the people in different occasion and situation they are facing and perhaps may come across.

In the same vein, Mawrie portrays nature not simply as the dwelling habitat that offer secured space for their life existence rather nature is but an institution that embrace diverse sort of life lesson necessary for the growth of life and daily experience (Mawrie, 1981). He further expounds this notion when he compares nature with a book containing ample informative and instructive life lesson that could help equip them to confront every life challenges that come along the way (Mawrie, 1981). The valuable learning process and directions align with their daily living is worth mentioning. The wisdom accumulated and guiding principles obtain from nature will help facilitating providing them important direction to address diverse issues affecting their daily life. At the same time it will also enrich their experiences in their journey towards a better future. It is believed that such a learning experience acquire from different signs and events offered by nature and surrounding environment is being considered as valuable ecological and natural treasure for the growth and development of life. From sowing the seeds to planting different sapling, from cutting trees and bamboo to making effort to preserve them nature provide them enormous lesson necessary for the growth and development of life. Some of the indigenous wisdom learnt from the nature around may be discussed below:

First, the best season for cutting timber for house building and other purposes to make them more lasting will be from the month of October till January before the trees bring forth their new shoots.

Second, it is a customarily practice by the people from rural areas to avoid using different tools made from bamboo and other canes like agricultural basket, winnowing basket, conical native basket, and low stool made of plaited canes and the like right after weaving them. Rather they will prefer keeping them for a year or more in a false ceiling makeshift mostly build in the kitchen for making them more lasting. Such makeshift ceiling meant for keeping these materials normally constructs in the kitchen with the intention to get expose from the warmth produced from the firewood smoke.

Third, sowing of seed or planting any agricultural saplings during the time of bright moonlight at night is generally avoided. This practice is done with the belief and understanding that this is the most preferred time for different insects to search for food.

Mother Nature speaks and communicate to human persons providing them some life lessons practically beneficial and helpful to different aspects of life. Commenting on the important life lesson and teachings imparting by Mother Nature to the people, in his poetry *Ka Meirilung Soso Tham* (1960) maintains:

> *Sa shisien pat kin win ki khlaw* (Once again the forest will roar)
> *Sa shisien pat kin khih ki maw* (Once again the rock will move)
> *Kum kiwei pat ki sngi kin mih* (Like other time the days will come anew)
> *Da kumwei pat ka Rik an ih* (In a new way the land will turn)
> *Lada don ki shkor ban sngew* (If we have ear to listen)
> *Aiu ka kren ka Mei-Ramew* (What the mother earth has to tell)

The above quotation illustrates that Mother Nature communicate to human person in various ways, helping them encounter life challenges in diverse situation. This learning experiences through their encounter and interaction with the natural surrounding generally shape their outlook and thought pattern. Nature is like an open book with numerous life affirming lesson offered to human persons in diverse space and time. For the indigenous people different components of nature whether living and non-living has certain important stories to tell to each generation that align with their growth and development. K.D. Ramsiej (1992) argues that all creations on the planet like the mountains, the rivers, the trees, the rocks etc. in their allocated abode not only situate there to exhibit the beauty and splendour of the natural surrounding but they have in one way or the other something valuable to inform and enlighten the people about important lesson related to the socio-cultural life of the people. It is these kinds of stories that offer important teaching and learning lessons that could help benefited different aspects of life and experience.

Indigenous People and Forest

Forest has become an important component closely attached to the life and experiences of the indigenous people. They play a fundamental role in the life and experiences of the indigenous people. Their long association with forest equipped them to have better knowledge about the dynamic and rhythm of the role and importance of forest to different lifeforms. It is this close relationship with the forest that practically inform them about the substantial value and extensive worth of the forest and forest cover. For many centuries back their ancestral understood the value of forest not only for providing them innumerable food items, fuel for cooking, medicinal stuffs, timber for house construction, rather for pure air and clean water. R.T. Rymbai (1998) argues that Khasi ancestral foresee the importance of forest not only for the present generation but also for the generations to come. That is not only for essential domestic necessity and other related services but for ecological services offered to all life forms. As he rightly puts it,

> There must, therefore, always be a sufficient green cover of trees for mother earth to maintain the balance of nature, in order to provide human with his/her basic necessities for sustaining life—not only wood for building and domestic use, but also for pure air and water (Rymbai, 1998).

It underscores the ancient knowledge and primeval wisdom of the indigenous ancestral anticipating the emerging needs and demands for forest resources and forest cover for both domestic requirement and ecological natural benefits and services. Such a profound information related to this ancient ecological treasure entails the need and obligation to preserve and protect these primeval resources since time immemorial. In other words, it demonstrates the closeness and attachment that connect the indigenous people with the forest that need to be maintained and preserved. One may note that forest and indigenous people cannot be separated. The moment they are separated it will have a negative impact to both sides. Pertaining this philosophy ancestral parents for many centuries back uphold this indigenous wisdom not only to preserve these ecological treasures but also to advocate for the need and requirements of generations to come. In this context, they categorically classified the forest cover into two categories—Sacred Forest and Prohibited Forest.

Sacred Forest

Sacred groves or the virgin forest are kept intact with the ultimate objective to maintain the balance of the mother earth and the surrounding environment. Some among different terms used for these types of forest include *Ki Law Lyngdoh, Ki Lawniam, Ki Law Kyntang* and many other local contextual terms. The idea of conserving the forest may be traced back to the pre-Vedic age (Ramakhrishna, 1998). Therefore, it may be perhaps as old as the indigenous community itself. The idea of preserving the sacred groves is closely associated to the social, religious and cultural life of the indigenous community. The presence of numerous monoliths within and nearby locations of these groves testify to that fact. According to the traditions, number of rites, rituals, religious ceremonies, taboos and the like that has been connected to these groves. Nevertheless, the ecological benefits and services rendered by these types of forest cover is meant not only for human persons and other life forms but for the entire planet. In other words, it is more of a cosmic implication rather than merely anthropic in nature. This depicts the ancient wisdom and vision of the ancestral parents to acknowledge the emerging needs of both the mother earth and other life forms. Some of the basic vital roles and functions rendered by these groves are:

First, as the entire vicinity within the parameter of the groves is considered sacred cutting any tree or even its branches, damaging any component, killing or hurting any life forms, hurting any ecological element within the domain and realm of sacred forest is strictly prohibited. They are supposed to be kept well intact without disturbance from any external forces that may distort their sanctity. Hence, in the midst of ecological damages in the surrounding environment, the presence of these sacred groves proved to be the safe haven for both living and non-living entities such as birds, animals, insects, vegetation, medicinal plants, and so on. Besides providing safe and secured habitat to different components they also serve as critical and strategic location for rich biodiversity and safeguarded home to ample species of rare and endangered plants and other categories of wild orchids and the like. Tiwari reported that sacred groves offer secured abode to "rare and endangered plant species and as many as 514 species representing 340 genera and 131 families were present in these forests" (Tiwari *et al.*, 1998).

Second, as many of these groves situated in areas that receive maximum amount of rainfall occur every year during summer, it is not wrong to say that they could protect land and soil erosion from being washed away by monsoon season. Being a virgin forest area, they could protect the land within their domain from all sorts of intrusion by natural and human made calamities.

Third, besides considered sacred groves by the community most of these groves situated in certain critical catchment areas having strategic importance. Many of these groves are located in specific and particular sites where the sources and origin of many major rivers and perennial streams found. Hence, it is not wrong to say that the groves not only are able to protect the critical sites of catchment areas but also could maintain the quality of water in the rivers and streams downhill. A classic example here is Shillong sacred groves on the top of Shillong Peak where the source of eight main rivers and streams originated (Tiwari *et al.*, 1998). Hence, the ecological services and valuable rich benefits contributed by these sacred forests to the local area and the natural environment is commendable needed to be cherished and fostered.

Prohibited Forests: Apart from sacred forest, indigenous Khasi people also have some other forest set aside for the need of the community in different circumstances. These forests are classified as prohibited forest that has been preserved since time immemorial. This category of forests may include *Ki Law Adong* (protected forest), *Ki Law Sang* (forbidden forest), *Ki Law Shnong* (village forest), *Ki Law Raid* (Raid forest) and many other forests in different region of the land. The basic difference here is that while in sacred forest cutting or plugging anything within the domain of the forest is forbidden, in the prohibited forest felling of trees and other forest resources is allowed with certain restriction. That is with certain restricted permission from the concerned authority entrusted to look after and maintain the forest. For instance, in case of *Law Shnong* cutting of trees for firewood or house construction is allowed to a certain extent, in other forests, like *Law Adong,* the same is allowed if the concerned family facing misfortune like natural calamities or other emergencies needed for those families (Rymbai, 1998). Hence, the concept of conserving the forest is done keeping in mind not only the need and well-being of the members of the community but also concerning the health and betterment essentially required to maintain the ecological balance of the mother earth. Indigenous ecological wisdom demonstrated the stark reality that the common welfare and betterment of human person is to be seen in connection to the health and well-being of the mother earth. They are very much part and parcel of the created whole rather than apart of other creations. Hence, it is not wrong to say, "The well-being of humans as well as other creations lies in the ability to maintain the rhythm of the mother earth" (Hayong, 2020).

Impact of the External Forces: Mirror of Today's Reality

Down through the centuries, indigenous people constantly confronted with numerous challenges that has severely affected their daily life and experiences including their relationship with nature around them. The coming of several external dominant forces like colonization, modernization, Christianization and of late globalization into the indigenous community brought tremendous impact to different aspects of their life and experiences. Indigenous people's attachment and deep relations to nature and Mother Earth gradually weakened down. Due to the impact of these external forces their relationship with land and forest has suffered to the certain extent in different capacities. The outcome is that perception towards nature and its very resources that has been fundamental source of living contributing enormous benefits to the native people is now changed. Nature and its resources are now looked upon as object or commodity to be exploited for economic value and maximum monetary benefit. The ecological value as well life subsistence significance of earth delicate but rich resources is being considered for their commercial profit of the present day generation ready for maximum exploitation within a short span of time. Their eco-based culture align with the rhythm and dynamic of nature and surrounding environment has been distorted and damage to the maximum extent. Indigenous cultural festivals and social activities in tune with the pace and pulse of nature slowly lost their essential connectivity and attachment to the beat of nature. For example, each and every indigenous cultural festival is normally grounded and performed in line with important natural events either with the sowing season or harvesting season. But today that alignment and association with natural happenings gradually diminished if not totally lost. This has not only distorted their eco-based cultural life experience but at the same time damaged their source of living with uncertain future at stake. Hence, in the midst of ecological imbalance and crisis that has threatened not only economic survival but the very existence of all life forms there is an urgent need to reconsider and rediscovery of their relationship with nature. Reconsider their beautiful and valuable relationship with nature here means to acknowledge and recognise their nature or eco-based life realities that may perhaps be beneficial towards the health and betterment of mother earth. At the same time, rediscover their nature based cultural life experience here means to retrieve and re-claim those cultural elements that has been lost and bring out their importance and reinterpret their significance to the present day society making them relevant to the changing need of time.

Indigenous Peoples' Ecological Resources: A Model for Environmental Preservation

In the midst of increasing ecological disaster that have threatened the survival of the planet earth's ecological resources of the indigenous people has something good to offer them may perhaps contributed to the betterment of the planet making it a better place for all creation. To state differently, if one has to confront the emerging challenges of the present day ecological crisis that has threatened life survival on this planet the value and worth of ecological resources as uphold and maintained by the indigenous communities need to be taken into consideration. In this attempt, drawing the insight from enormous value and importance of ecological resources as perceived by the indigenous people is not only necessary but mandatory. In the midst of impending uncertain ecological crisis, some of the important insights drawn from these resources could be served as model of environmental preservation.

Value of Land: Reclaiming the Indigenous Perspective

For the indigenous people land is not merely a space to live but a place where the entire life existence finds its meaning. The space one's grow up, the landscape that shape a person's attitude, the difficult terrain that brought enormous life experiences, the local native surrounding that contribute in ample ways towards the growth of the individual personalities and the like, all these together give land a distinct value that need to be acknowledged, recognized and cherished. Some of the important components that portray the value of land while at the same time could serve as the model to preserve land from further exploitation has been discussed below:

First, for the indigenous people land is the most precious entity inherited and handed down from generations to another. The creation story of the Khasi people (Hayong, 2008) tells that land is the creation of God and the gift of God to the people. Being the gift of God land do not belong to the people rather people belong to the land. In other words, we do not own the land but the land own us. We are very much part and parcel of land and not apart from land. We inherit it from our ancestral and expected to secure its well-being transferring it down to the future generations. If land do not belong to us but we belong to land, it is in reality, implies that humans have no right to exploit or damage it but to maintain it. Such important perception related to land enables the indigenous people to pay deep respect to land since time immemorial. It underscores the necessity to protect and conserve it from any sort of distortion and exploitation. There is a Khasi saying related to land, which says *Ka Ri Umsnam u Ni u Kong* which literary means land is a hard-earned space and our ancestral achieved it with great sacrifices while at the same time sustaining the same for the essential needs and requirements of future generations. Hence, land for them is the most treasured constituent necessary for the safety and security of the community not only for the previous generations but for many generations to come. It demonstrates the importance of land that could bind together connecting the past history of the community with the future life reality. Such a concept conveys that there is no room for damaging or exploiting but more opportunities of conserving and preserving the value of land. Hence, if the ancestral parents understand well the value and importance of land that need to be preserved it may perhaps offer crucial lesson to the present day generations to acknowledge and recognize the same.

Second, originally land being God-given is a precious entity meant to be shared not for sale. In the past, people did not sell the land but shared it. This is practiced with the understanding that ecological value of land is more important than the economic or commercial monetary value. In other words, the ecological services offered by land display more value and worth than the commercial monetary value bargaining for few financial benefits. According to the land tenure, system of the Khasi people land is classified into two categories—*Ri Kynti* and *Ri Raid* (Hayong, 2018). The former belongs to the family or clan inherited from their ancestral where all the members of the clan have the inherent right to share a portion of it. Whereas the latter belong to the community and all the members are entitled to have a part of it for their needs either for dwelling place or for agricultural related activities. If one referred to the traditional belief system of the Khasi people selling of land especially *Ka Nongtymmen* or the inherited land is strictly prohibited. If there is a need to do as such, land has to be sold to the members of the clan not outside the clan circle (Hayong, 2018). It denotes their understanding the value and importance of land that need to be kept intact without any distortion.

One may note that prior to the coming of British rule selling of land is unknown by Khasi people. For instance, when the British administration required land to establish their administrative centre, the Syiem of Sohra that time did not sell land to them but provided them the land in exchange with the land handed down by the Britishers in Sylhet area (Giri, 1998). It denotes that till the coming of British people selling of land was not an issue to the native people. But this proved otherwise in the present day ground reality. The impact of external forces gradually changed the native people's perception of land. Today the value of land is compromised to short-term monetary financial benefits at the expense of life-long ecological services that land may offer to the entire creations. Basing the value of land for economic gain demonstrates that the more a person acquired land the more his or her status and respect establish in the community. It openes up the window for the powerful and the rich tycoons to exploit the situation for acquiring or accumulating more land at the expense of the poor and needy. Hence, the basic and original concept of land as a sharing entity is being replaced by commercial economic exploitation benefiting the selected few rich section of population. Hence, there is a gradual shift from sharing model to acquiring model of land accumulation leaving majority of people remain landless.

According to the report, available landless section of population in Meghalaya today is 76 per cent of the total population (Momin, 2020). The report shows that accumulation of land is at the hand of few selected privilege section, leaving majority of population remain landless. Besides the price rocketing of land of commercial business related matters is beyond the reach of many poor section of population. Hence, it entails the need of strong legal protection such as Land Ceiling Act so that the model of land redistribution may somewhat bring certain solution to the emerging problems and challenges confronted by the present day community. Having said that it does not mean to go back to the past or to say no to the issue of land selling rather it demands stringent action plan making land availability to different section of population. In addition amidst certain emerging changes related to the concept of land one must not forget the ecological value offered by land making the inhabited earth a better place to live in.

Third, recognizing the value of land for ecological preservation necessitates to rediscover the indigenous people's concept of motherhood of land. Land for the indigenous people is like their mother that gives and nurtures life. The concept of earth as the mother stands at the hearth of indigenous people relations to the earth. Earth as a mother could encompass and embrace all life forms in its bosom with adequate room to grow, to flourish and to be fruitful. Hence, the three conceptual usage for mother earth as *Ka Mei Mariang, Ka Mei Ramew, Ka Meirilung ka Meirisan* depicts an in-depth notion of the close intimacy and respect of the earth as their mother. It reflects the intrinsic value inherently inscribed in different layers and features of mother earth that could help nurturing, safeguarding and promoting all life forms amidst diverse challenges. But the entry of diverse major upheaval forces affecting different aspects of life such beautiful and meaningful insights of the native people began to diminish if not totally lost. Moreover, in the midst of ecological crisis and looming imminent threat jeopardizing the flow of life and survival reclaiming the concept of the earth as a mother could become a powerful and prevailing model for conserving the planet from further crisis. Reclaiming the concept of mother earth (Hayong, 2020) does not mean to go back to the primeval or primordial distant past experiences where the earth was totally free from all sort of exploitation and discrimination for such an idea may seem irrational or impractically possible. Rather reclaiming mother earth means to stand against all policy matters, resisted all major forces that damaged the flow of life and resisted all attempt to exploit the earth and its resources for temporary immediate economic benefit at the expense of future generations. For the indigenous people it surely means to return to the basic of their relationship with the earth, to find out their roots that connect them with the mother earth, making all efforts to regenerate the sustaining power of the mother earth that could sustain life and once again transforming it into a better place— a secured home for all things that exist.

Reimagining the Concept of Forest: Indigenous Ecological Model for Environmental Preservation

Forest play very important role in the life and experiences of the indigenous people. It may be not wrong to say that among various communities, indigenous people have better knowledge and understanding concerning the value and other things related to forest. For them, forest value is associated not only on things related to human essential needs but also to crucial ecological services required for both living and non-living entities. The two categories of forest, *Law Adong* and *Law Kyntang* are being preserved for the needs of human in addition to crucial ecological benefits for the entire planet. This portrays the distinct ecological wisdom of the ancestral parents to foresee the value of forest not only for human essential requirements but also for the natural services rendered by the forest that could help maintaining the balance of the eco-system. In the midst of threatening ecological tragedy, ingrained divinely ecological wisdom of the indigenous ancestral parents could offer certain possibilities to solve the impending threat of ecological crisis.

Law Adong Model

The underlying idea behind instituting important constituents like *Law Adong* is but to provide the members of the community equal access to all essential requirement needed from the rich forest resources. It is interesting to note that all members have the privilege to get access and acquire forest resources necessary with limitation in accordance to their need not more. This is normally practice keeping in mind to acknowledge and recognize the value and worth of each and every resources within the vicinity of the forest. The important rule related to this are:

First, all members have the freedom to equally share the basic resources but use them for meaningful purpose not otherwise. To state differently, they are supposed to use but not misuse these valuable ecological resources.

Second, together with the equal right accessible concerning these resources, all members are expected to use exclusively what is needed with certain amount of limitation. This implies that they are not expected to procure more than necessary. Hence, it is need based not greed based. The idea behind such beautiful concept is to keep maintaining these valuable treasures so that they can sustain for many generations to come.

This beautiful perception illustrates the ecological wisdom of the indigenous ancestral parents relating to judiciously use the forest resources intent to be earmarked not only for the present day society but also for many generations to come. This striking and meaningful example could be a significant model for today's generations in our attempt for environment preservation. For them, forest and its resources are intended not only to meet the need and requirement of the present day society but for long-term sustainable benefits of the community.

Sacred Groves Model

Like other reserved forests, sacred groves have become part and parcel of the indigenous peoples' culture and experiences. In the present day context where ecological crisis brought impending and serious threat to the entire planet and ecological system, sacred groves offer a viable relevant model to save the mother earth from the further ecological crisis. It is not wrong to state that sacred groves could provide essential ecological services both within and outside its surrounding.

First, within the sacred groves, all creations both living and non-living could enjoy a secured home and safe dwelling habitat without no hindrance that could interrupt the overall well-being and welfare. There is no forces that could hinder their growth and progress in their respective allotted space within the realm of sacred groves. In the present day ecological debate a new way of looking at the earth emerged. That is, the whole natural environment is not only a place where one can share their habitat but it is a secured home to all creations. Therefore, each component either living or non-living have their respective role and responsibility to its health and well-being. Commenting on the new way of looking at the earth as home, M.P. Joseph argues that home is a place that sustains life, allows its members to blossom freely, welcomes new life and foster the continuing flow of life. Moreover, home is a place that empower and strengthen the bond of relationship of all the members without allowing anyone to feel like a stranger (Joseph, 2009).

Second, in the present day era where ecological crisis has disturbed all life forms, the concept of harmonious relationship need to be re-defined. Indigenous community and their worldview depicts a perfect example where harmonious relationship among all created things is well maintained. Their folklores, folktales, myths, legends, stories possess important insights that portrays that all living creatures could live together in peace and harmony. In other words, all creatures either big or small could live together in a peaceful atmosphere in unity and harmony. The spirit of harmonious relationship is well maintained within the parameter of sacred groves. Each and every creation both living and non-living have their own dignity and respect. The entire ground and earth resources are free from any sort of exploitation, distortion and damage from any forces that may hamper their growth and well-being. Such an ambient serene atmosphere where dignity and respect are maintained resulted to ecological balance of the entire cosmic system. If human beings could realize the value and dignity of all the created order they will not misuse them for their selfish gain but utilize them with utmost care and concern. If one has to strive forward towards a harmonious relationship of all creations, the value and worth of each created things need to be acknowledged and recognized. It is, therefore, a call to re-discover and "affirm our interrelatedness with nature and commit ourselves to honour the integrity of creation" (Abraham, 1993). Reconsider our relationship with other creations is nothing but to realize our duty and responsibility towards them in particular and the entire inhabited earth in general.

Third, an understanding of the importance of harmonious relationship give way to a new way of discovery of the inherent system of conservation. This is clearly elaborate in the concept and practice of keeping the sacred groves without any disturbance. One may note that while the area surrounded the sacred groves always laid barren and desolate scarcely having green cover, the sacred groves stand tall as a living symbol of conserving the bio-diversity. Indigenous people understand that the well-being of humans depends largely on the well-being of other creatures and the health and well-being of mother earth. Hence, they made maximum efforts to preserve the entire surrounding environment because of the belief that all created things depend on each other and not otherwise. Commenting on the need to preserve the natural environment and its resources Samuel Ryan states "to live and act is to breathe the air, to drink some water, to eat fruits, roots, seeds and leaves etc. We depend on the mineral, plants, animals, these depend on a thousand cosmic, physical, chemical, magnetic factors, forces processes and interactions" (Ryan, 1991). It depicts the importance and worth of each and every created thing needed for the balance of the eco-system. They are meant to be used not to misuse, they are supposed to be preserved not for the sake that they are useful to human beings, rather because they contain the capability to contribute towards the health and well-being of the ecosystem.

Conclusion

By indigenous people ecological resources have been preserved intact down through the centuries for the need and well-being of humans as well as natural environment. However, the entry of dominant external forces resulted to exploitation and distortion of these precious resources leaving the entire surrounding environment in peril. Many of these resources have been exhausted, so also the ecological services they basically offer has been diminished. This entails the need to re-discover and re-consider their value and ecological insights and re-interpret them from the lens of the present day context. In such an attempt, new ecological insights could be discovered making them useful and meaningful resources that could help bring important solution to the present day ecological crisis.

REFERENCES

Abraham, K.C. (1993). "Theological Response to Ecological Crisis". *Bangalore Theological Forum,* XXV/1 (March, 193): 13.

Chowdhury, J.N. (1996). *Ki Khun Khasi Khara* (*The Khasi People*). Shillong: Published by the Author.

Giri, Helen (1998).*The Khasis Under the British Rule, 1824-1947*. New Delhi: Regency Publications.

Hayong, S. (2020). "Adivasi Eco-Diakonia: A Theological Probing from the Khasi Perspective". *Adivasi Journal of Theology,* 2/2 (October, 2020): 52-53.

_______(2008). "The Khasi Concept of Ecology". *Indian Theological Journal,* 2/1 (January-June 2008): 87-88.

_______(2018). *Tribal People and Globalization: A Theological Perspective*. New Delhi: Christian World Imprints.

Joseph, M.P. (2009). "Where is your Sister/Brother? Carbon Civilization defiles Faith in God". In Thomas Samuel and Mathew Koshy Punnakadu (eds.). *A Christian Response to Ecological Crisis.* Thiruvalla: Christava Sahitya Samithy.

Kozhamthadam, S.J. "Forward". In Sarojini Henry. *We're one Earth Community*. Delhi: ISPCK.

Longchar, Wati (2012). "Liberation Theology and Indigenous People", in Hrangthanchungi, M.M. Ekka and Wati Longchar (eds.). *Doing Indigenous Theology in Asia: Towards New Frontiers*. Kolkata: NCCI/GTC/SCEPTRE.

Mawrie, H.O. (1997). *Ka Pyrkhat u Khasi*. Guwahati: Silpi Printers.

_______(1991). *Ka Theology Jong ka Niam Khasi*. Shillong: Ri Khasi Press.

_______(1981). *The Khasi Millieu*. New Delhi: Concept Publishing Company.

Minz, Nirmal (1998). "Tribal Perspectives on Ecology". In A. Wati Longchar and Yangkahao Vashum (eds.). *The Tribal Worldview and Ecology*. Jorhat: Tribal Study Centre.

Momin, Jakrimre and Bhogtogram Mawroh (2020). *Land Tenure System and Landlessness in Meghalaya: A Literature Review*. Shillong: NESFAS.

Nongkynrih, Kynpham Sing (2007). *Around the Hearth: Khasi Legends*. New Delhi: Penguin Books.

Ramakhrishna, P.S. (1998). "Conserving the Sacred for Bio-diversity: The Conceptual Framework". In P.S. Ramakhrishna, Saxena, K.G. and Chandrashekara, U.M. (eds.). *Conserving the Sacred*. New Delhi: Oxford/IBH.

Ramsiej, K.D. (1992). *Ka Mariang ha u Khasi*. Shillong: Bonsen Press.

Ryan, Samuel (1991). "Ecology and Faith: The Mystery of the Earth". *National Council of Churches Review* CXI (October 1991): 1281.

Rymbai, R.T. (1998). "The Traditional Ecological Concepts of the Khasi-Pnars". In A. Wati Longchar and Yangkahao Vashum (eds.). *The Tribal Worldview and Ecology*. Jorhat: Tribal Study Centre.

Tham, Soso (1960). *Ki Sngi Barim u Hynniew Trep*. Shillong: Ri Khasi Press.

Thanzauva, K. (2004). *Theology of Community*. Bangalore: Asian Trading Corporation.

Tiwari, B.K., S.K. Barik and R.S. Tripathi (1998). "Sacred Groves in Meghalaya". In P.S. Ramakhrishna, Saxena, K.G. and Chandrashekara, U.M. (eds.). *Conserving the Sacred*. New Delhi: Oxford/IBH.

War, Juanita (1998). "The Khasi Concept of Nature". In A. Wati Longchar and Yangkahao Vashum (eds.). *The Tribal Worldview and Ecology*. Jorhat: The Tribal Study Centre.

9

Indigenous Practices of Environmental Governance in the North East India

A Brief Review

Ashish Malhotra

Introduction

Over the ages, indigenous people have developed innumerable technologies and art forms. They have devised ways to farm deserts without irrigation and produce abundance from the rain forest without destroying the delicate balance that maintains the ecosystem, they have explored the medicinal properties of plants, and they have acquired an understanding of the basic ecology of the flora and fauna. Much of this expertise and wisdom has already disappeared, and if neglected, most of the remaining could be extinct in the near future.

Until quite recently, very few in the developed world cared much about this cultural holocaust. The prevailing attitude has been that the so-called Western Science, with its powerful analytical tools, has little to learn from tribal knowledge. The developed world's disastrous mismanagement of the environment has somewhat humbled this arrogance. However, some scientists are beginning to recognize that the world is losing an enormous amount of basic research, with indigenous people drifting away from their culture and traditions. Scientists may someday be struggling to reconstruct this body of wisdom to secure the developed world's future.

Human impact on nature has reached such a high proportion that the world is today witnessing an extraordinary rate of species loss. In 1988, the International Union for Conservation of Nature and Natural Resources (IUCN) listed 4,589 threatened animals. By the year 2050, at least 60,000 plant species will become extinct or threatened. According to a report of biologists, the current rate of extinction is at least 25,000 times greater than the intensity of extinction that took place during evolutionary times. The rate of extinction of mammals alone has risen from one species every five years in the 17th century to one in every two years in the 20th century.

There are many records in the form of written or unwritten traditional lore that, traditional societies have a very good concept of ecosystem management. Indigenous people have a logic of life, which is very different from that of modern man. It is because of this logic that indigenous people are where they have been for ages and not because of their backwardness or primitiveness. These people decided not to run for so-called development. Once one runs for development one does not know when to stop and does not know why one is running, it is hard to define what development is.

Sacred groves, sacred ponds, sacred patches of grasslands, sacred animals, and others are examples of traditions of conservation backed by religious sanctions. The preservation of biological resources by such traditions is of immense significance. Certain species of plants and animals have been preserved on account of sacred qualities attributed to them. The most common examples of such trees are those of *Ficus religiosa*, *neem*, mango, wood apple etc.

Plants and animals worshipped as totems symbolize the kinship ties of humans and nature. These totems, sacred to specific clans, are accorded full protection. This practice is mainly prevalent among the tribal populations and especially among those who undertake hunting and gathering in some form or the other. The practice of certain taboos with regard to resource extraction could be interpreted as conservation practices cast in religious idioms, as most of them allowed for regeneration and perpetuation of species.

In addition to the belief systems, which condition the relations between humans and biodiversity, the social, political and economic relations between the humans themselves influence or control the use of resources. Thus, the indigenous community and wider institutional structures form the second context of resource use. These include structures for the management of common property resources, customary tenure rights, customary laws and rules, regarding resources, localized economics and so on. Traditional community based management of common property resources typically imposed restrictions on the indiscriminate use of resources and ensures some form of distribution of benefits and livelihood opportunities. The self-managed village commons (water bodies, forests, pastures etc.) were often equitably managed.

The diverse types of forests in India are perhaps the richest assemblage of biodiversity. The dependence of traditional communities on these forest resources is enormous. Being of such critical socio-economic importance, these communities tended to exercise self-restraints in the use of forest resources, thus conserving both the constituent biodiversity of forests as well as diverse forest type as a whole. Beside these communities are also vast repositories of forest related biodiversity knowledge.

Environment Management in the Religious and Cultural Heritage of India

An in-depth and unbiased investigation of the religious and cultural heritage of Indian society reveals that they have, over centuries, provided knowledge of nature and rules of natural resource utilization that has enabled ancient civilizations to survive materially. The fundamental constituents of the material world are defined in these systems in a manner, which respects the integrity of nature and the stability of essential ecological processes. India's religious and cultural traditions are rich and diverse which includes a rich resource of beliefs about nature and rules for the sustainable utilization of material resources.

Plantation of trees in the public places, such as roadsides, temple gardens etc. is an ancient tradition of India. The abundant age-old trees found throughout the holy shrines in India reveals an age-old eco-friendly Indian tradition. Trees such as mango, *neem* or banyan have always been planted along the roads to give shelter and shade, their leaves acting as natural air-conditioners. These huge trees have, for generations, proved to be comforting shady haunts for weary travellers.

Planting trees and digging wells have traditionally been the two great acts of charity by which anyone could earn merit and universal appreciation. This tradition of valuing trees was passed on in to Indian culture and led to a subtle ecological relationship between human communities and the forest community of trees, plants and animals.

Concern for the welfare of the natural world has been an important element throughout the history of Buddhism. The recognition that human beings are essentially dependent upon and interconnected with their environment has given rise to an instinctive respect for nature. Although Buddhists believe human beings have a unique opportunity to gain enlightenment, which other creatures do not, they have never believed humanity is superior to the rest of the natural world (Batchelor, 1992).

Buddhism advocates a gentle, non-aggressive attitude towards nature. According to Buddhist philosophy, a householder should accumulate wealth, as a bee collects nectar from a flower. The bee harms neither the fragrance nor the beauty of the flower, but gathers nectar to turn it into sweet honey. Similarly, a human being is expected to make legitimate use of nature so that he can rise above nature and realize his innate spiritual potential (De Silva, 1991).

Among Buddhists, large, old trees are particularly revered. The attitude, which is a legacy of pre-Buddhist animism, does not violate the belief system of Buddhism. The trees are called *Vanaspati* in Pali, meaning 'lords of the forests'. The deference to trees is further strengthened by the fact that huge trees such as the ironwood, the *sala* (*Shorea robusta*) and the fig tree are acknowledged as Bodhi trees, trees under which Buddha attained enlightenment. It is well known that the fig species (*Ficus religiosa*) is held as an object of great veneration in the Buddhist world as the tree under which Buddha attained enlightenment (De Silva, 1991).

The great ancient Indian king Ashoka had passed the famous abstinence for protection of many wildlife. In his famous abstinence ordinance, the following animals are declared inviolable—*Suka* (Parrot), *Salika* (Myna), *Cakravaka* (Ruddy goose), Hamsa (Wild goose), *Nandimukha* (a kind of bird), *Gelataor Jetuka* (Bat), *Ambakapilika* (Queen ants), *Dali* (Terrapin), *Anathika maccha* (Jellyfish), *Vedaveyaka. Gamgapuputaka, Samkujamaccha* (Skate-fish), *Kaphatasayaka* (Porcupine), *Pamnasasa* (Squirrel), *Simaqla Samdaka* (Wild bull), *Akapinda* (Iguana), *Palasata* (Rhino), *Seta kapota* (White dove), *Gama kapota* (Domestic dove), he adds the significant clause "which is neither useful nor edible". That the spirit of the edict is not less economic than altruistic is further proved by the forest law—"forests must not be burnt either uselessly or in order to destroy living animals" (Raichoudhury, 1964).

The preservation of the four-footed, feathered and finny races is sought with assiduous care in other rules of the economist. For this specific purpose the *Abhayaranya* (Sanctuary) is set apart and none are allowed to entrap, kill or molest deer, birds and beasts protected thereunder (Bose,1945).

Traditional Environment Management in North East India

North East India is one of the rich areas of biodiversity. It is an established fact that the areas which are rich in cultural diversity are also rich in biodiversity. It is also a region rich in cultural diversity. People of North East India preserved the nature and natural resources of this part of the country as a part of their traditional culture. But, as the indigenous peoples of this region has been threatened for a long time, as development encroaches on their lands and traditions, the Biodiversity of this part of India is also greatly threatened. Therefore, an effort for preservation of traditional culture for the sake of conservation of nature and natural resources is the need of the hour. For a better understanding of the significance of traditional practices for conservation and management of environment in the present day context, a study in this line is of utmost importance. Various ethnic groups of people of North East India have practiced a large number of traditional systems of environment management. Some of these practices are described below:

Traditional Rainwater Harvesting

Collective action is the most important aspect of traditional people's "management of environment" in North East India. Dugout-*cum*-embankment type of water harvesting structures were used for storing rain or runoff water traditionally by most of the tribes and communities of this region. Initially, the main purpose of water harvesting was to meet drinking water requirement of humans and animals during scarcity. With the development of the agricultural system, the utility of such water storages began to be extended to irrigation. People of North East India retain and harvest water in several ways.

Several hundred years ago, tribes in Meghalaya developed procedures of harvesting water from hill streams from bamboo pipes for irrigating betel vines and other plantations on steep slopes. With a network of bamboo pipes of different diameters, lengths and shapes, 18 to 20 litres of water were diverted per minute from the main channel to flow at the rate of 20 to 80 drops per minute at irrigation points. The beauty of the system lies in the use of local materials and resources, and the perfection in water distribution without leakage. Streams are also diverted for irrigation with the help of channels or pipes.

Traditional water harvesting units were managed and maintained by rules and procedures commonly agreed upon by the villagers. Such rules and regulations are still prevalent in villages of North East India, particularly in the hills. Problems relating to water management are not the result of any single individual's activity, rather the cumulative effect of the actions of many individuals. Traditional systems, if studied and evaluated in totality and not just for their technology, can provide some answers to the current problems in our environment (Singh, 1989).

Rainwater Harvesting in Assam

Assam has a long tradition of rainwater harvesting by different methods. The Bodo tribe of Assam is one of the premier ethnic groups of people to start wet cultivation of rice in the earlier Assam. The cultivators of Assam, particularly in the areas of

Bodo inhabitants usually never face a drought condition in their rice fields, as there are age-old man-made canal systems, which canalize rain and spring water from the distant hill springs or lakes. In Assam, this age-old man-made canal system is called dong. The dongs are though not very wide but very long, distributing its branches and sub-diverted throughout an extensive paddy field. It is estimated that a single dong may be capable of irrigating one thousand to five thousand hectares of land in the dry season (*Imperial Gazetteers of India,* Provincial series, Eastern Bengal and Assam, 1909). It is worthy to be mentioned that such a large and scientific irrigation canal system was the creation of the wisdom of the people. The villagers collectively made such a large irrigation system. Again, it is interesting to note that, there was standing rules for management and equal distribution of water to all the villagers at the time of need.

At the time when the water is necessary for drenching of a field, the farmers make a temporary mud bundh over the canal. Upon the completion of the bundh, the hydrological dynamics of the canal's upper segment will result in the inundation of the entirely of the canal situated above the bundh. At this stage, the longitudinal embankment of the canal has cut in places so that the water can enter the field. After complete inundation of the field to the desired level, the bundh is demolished.

"Pukhuri" (Tank)—The Marvels of Water Conservation in Assam

Construction of tank for conservation of water is the most ancient and glorified technology of water management in Assam. Tanks are small reservoirs built by constructing dams and it is the oldest method of water conservation by storage of runoff water. There are large number of tanks in Assam most of which were constructed in early times. The rulers of ancient Assam were much conscious about the construction of water tanks for public service. There were special royal officers to look after the construction of water tanks, especially in the days of Ahoms (Basu, 1970).

Construction of pond or tank either for domestic use or for public use is a worthy tradition of Assam. There is a strong belief that one can earn much God's grace, if he can construct water tanks or ponds for the public. Due to this traditional belief in the Assamese society, the rulers of Assam had an aim of constructing as many tanks as he could. The Ahom kings of Assam constructed a large number of large sized water tanks. Construction of tanks was a point of consideration about the greatness of the king. Most tanks were constructed for religious purposes, but these tanks served economic and agricultural needs. Tanks were constructed when a deity was being installed or on the demise of a loved one. There are many instances in folk tradition when women and men sacrificed their lives as well as those of their children in the belief that their sacrifice will bring water.

Traditional Water Harvesting System in Arunachal Pradesh

People of Arunachal Pradesh have developed various ways for harvesting of water in the hills and valleys. They not only strategically harvest rainwater but also use techniques for conservation of moisture and nutrients.

There are two important traditional irrigation systems in Arunachal Pradesh—the irrigation of rice terraces with the help of bamboo pipes and the *Apatani* system of wet rice irrigation. In the former case, water is transported through an intricate system of bamboo pipes to agricultural fields. However, this system is now becoming obsolete and is being replaced by iron pipes and channel irrigation.

The *Apatanis* have evolved a very scientific system of field irrigation. The striking features are the partially flooded rice fields, and the intricate design of the contour dams dividing the plots. The valley floor has a gentle gradient and the terraced holdings are laid out along the general slope. 0.6 m high earthen dams supported by bamboo frames divide the plots. All holdings have an inlet facing the water-head and an outlet on the opposite side. The inlets for the low-lying plots act as outlets for the higher level plots. A deeper conduit channel connects the inlet point with the outlet point. When a terrace is to be filled with water, the outlet is blocked. By opening and blocking the connecting ducts, any field can be flooded or drained as required (Malley, 1907).

Bamboo Drip Irrigation in Meghalaya

In Meghalaya, an ingenious system of tapping of stream and spring water by using bamboo pipes to irrigate plantations is widely prevalent and so perfected that about 18-20 litres of water entering the bamboo pipe system per minute gets transported over several hundred metres and finally gets reduced to 20-80 drops per minute at the site of the plant. The tribal farmers of Khasi and Jaintia hills use the 200-year-old system.

The bamboo drip irrigation system is normally used to irrigate the betel leaf or black pepper crops planted in areca nut orchards or in mixed orchards. Bamboo pipes are used to divert perennial springs on the hilltops to the lower reaches by gravity. The channel sections, made of bamboo, divert and convey water to the plot site where it is distributed without leakage into benches, again made and laid out with different forms of bamboo pipes. Manipulating the intake pipe positions also controls the flow of water into the lateral pipes. Reduced channel sections and diversion units are used at the last stage of water application. The last channel section enables the water to be dropped near the roots of the plant. Bamboos of varying diameters are used for laying the channels. About a third of the outer casing in length and inter-nodes of bamboo pieces, have to be removed while fabricating the system. Later, the bamboo channel is smoothened by using a *dao*, a type of local axes which is a round chisel fitted with a long handle. Other components are small pipes and channels of varying sizes used for diversion and distribution of water from the main channel. About four to five stages of distribution are involved from the point of the water diversion to the application point.

The water for betel leaf plants is diverted from streams by temporary diversions into very intricate bamboo canal systems. Betel leaf is planted in March before the monsoon. It is during winter that irrigation water is required, and the bamboo pipe system is used. Hence, these bamboo systems are made ready before the onset of the winter, and during the monsoon, no water is diverted into them.

The farmers themselves do maintenance of the pipes and supports. A cooperative has been formed, and each farmer provides his skill and labour to maintain the system. Repair work is undertaken as and when required. Diverting water from one field to another at fixed timings carries out distribution of water. To divert the water, a short bamboo with a hole at the bottom is placed across the main lines. This blocks the main water pipe and diverts the water.

Environment Management in Oral Lore

From early times, conservation of surface water by constructing very big ponds in the field has been a tradition in Assam, as well as in some other parts of the North Eastern states. During the days of Ahom kings, there was a very good tradition of constructing big water tanks, which are often regarded as Sager or sea. These water tanks never dried-up and were used as a reliable source of water in severe drought. The ponds were used simultaneously for field irrigation, for drinking water purpose and as fisheries. There is a very popular proverb regarding the conservation of water in Assamese:

> *"As you regard your own religion, Keep water like a queen constructing a pond."*

In autumn and early winter, the water in the rice fields starts drying up. However, this is peak time for the flowering and maturing of rice. During this time if the rice field dries up, it hampers the production of rice. Therefore, there is a tradition of keeping water in the rice field during the flowering time of rice in the autumn. In this regard, there is a very good proverb in Assamese:

> *"Conserve water in Kati and Aswin, as the king keeps the queen."*

Cultivation in the hilly slopes is a problem for want of waters, because the water flows down immediately to the plains. Rice is the manifold crop of the Assamese people. Moreover, wet cultivation is the most popular practice of the people. The "sali" variety of rice is the only one variety popularly used by the people. Cultivation of this variety needs a considerable amount of water in the field. However, in the hilly areas cultivators of this region developed a system of constructing embankments densely to conserve rainwater in hill slopes. There is an old proverb regarding this.

> *"Construct dense embankments on the hill, then only plant rice in the field."*

Good crop usually come from lands whose middle portion is low-lying. The reason why it is made fertile is the accumulation of water in the centre and flowing down of the fertile top soil from the higher ridges around. This is mentioned in a proverb:

"Buy a land whose middle is low, marry a girl whose mother is good, you know."

REFERENCES

Alleri, B.C. (1904). *Gazetteer of Bengal of North East India*. Delhi: Mittal Publications.

Arvil (1978). *Man, and Environment*. London: Penguin Books Ltd.

Banwari (1991). *Hindu Principles of Protecting and Organizing the Environment*. New Delhi: Ganga Kaveri Publishing House.

Baruah, B.K. (1961). *Assamar Lok Sanskriti, Assamese Folk Culture*. Guwahati: Assam Book Depot.

Basu, N.K. (1970). *Assam in the Age of the Ahom*. Calcutta: Sanskrit Book Depot.

Batchelor, M. (1992). *Buddhism and Ecology*. New Delhi: Motilal Banarsidass Publishers.

Bose, A.N. (1961). *Rural Economy in Ancient India*. Calcutta: Firma K. L. Mukhopadhyay.

Choudhury, P.C. (1959). *The History of Civilization of the People of Assam to the Twelfth Century A.D.* Guwahati: Department of Historical and Antiquarian Studies (DHAS), Assam.

De Silva, L. (1991). *The Hills Wherein My Soul Delights*. New Delhi: Indian Council for Cultural Relations (ICCR).

Elwin, V. (1958). *Myths of the North-East Frontier of India*. Shillong: North-East Frontier Agency (NEFA).

Gogoi, L. (1994). *Lila Gogoi Rachanavali* (*Collected Works of Lila Gogoi*). Dibrugarh: Purbanchal Prakashan.

Gray, A. (1990). *The Impact of Biodiversity Conservation on Indigenous Peoples*. London: International Institute for Environment and Development (IIED).

Gurdon, P.R.T. (1914). *The Khasis*. London: Macmillan and Co.

Hodson, T.C. (1908). *The Meitheis*. London: Macmillan and Co.

______(1911). *The Naga Tribes of Manipur*. London: Macmillan and Co.

Kakati, B.K. (1967). *Mother Goddess Kamakhya*. Guwahati: Publication Board, Assam.

Kothari, A. & Kothari, M. (1995). *Sacrificing Our Future: The New Economic Policy and the Environment*. Calcutta: Consumer Unity & Trust Society (CUTS).

Linden, E. (1991). "Lost Tribes, Lost Knowledge." *Time*, *138*(13), 46-56.

Malley, O. (1907). *Bengal District Gazetteers: Darjeeling.* Calcutta: Bengal Secretariat Book Depot.

Mills, J.P. (1922). *The Lotha Nagas*. London: Oxford University Press.

________(1937). *The Rengma Nagas*. London: Oxford University Press.

Mishra, P.K. (1990). *Ecological Stability in Tribal Lifestyle.* In S.C. Kalwar (Ed.), *India's Environmental Crisis and Responses* Cpp. 101-112). Dehradun: Indus Publishing Company.

Narzary, B. (1985). *Baro Kacharir Samaj Aru Sanskriti* (*Society and Culture of the Bodo-Kacharis*). Guwahati: Bina Library.

Playfair, A. (1909). *The Garos.* London: David Nutt.

Prime, R. (1992). *Hinduism and Ecology*. New Delhi: Deep & Deep Publications.

Raichoudhury, S.P. (1964). *Agriculture in Ancient India.* New Delhi: Indian Council of Agricultural Research (ICAR).

Ray, J.C. (1948). *Life in Ancient India.* Calcutta: Punthi Pustak.

Schumacher, E.F. (1973). *Small is Beautiful: A Study of Economics as if People Mattered.* London: Blond & Briggs.

Singh, A. (1989). *Bamboo Drip Irrigation.* Shillong: Indian Council of Agricultural Research (ICAR), North Eastern Region.

Thangsanga, R. (1987). *Approach to Increase Rice Production in Mizoram.* New Delhi: Centre for Science and Environment (CSE).

Tiwari, S. (1987). *Environment Crisis and Hindu Religion.* New Delhi: Ganga Kaveri Publishing House.

10

A Comparative Study of Forest Cover and Land Use and Land Cover Change in and around Two Sacred Groves in East Khasi Hills District, Meghalaya from 1990 to 2020 Using Landsat Imagery

Bennathaniel H. Diengdoh and *S. Jayakumar*

Introduction

Sacred groves are community-managed forests that merge traditional beliefs, attitudes and utilization. Since their management is inextricably intertwined with traditional belief and custom (Khan, Khumbongmayum and Tripathi, 2008), the perpetuation of these groves is often contingent upon the attitudes, perceptions, beliefs and activities of local communities that have instituted them. Among these activities are various regimes related to forest resource extraction and land use. The impact of wood extraction, for instance, may be reflected in a change in forest cover, while other forms of land use such as agriculture, construction and settlement expansion can be monitored through changes in land use and land cover using remotely sensed data. The use of data from different time periods can enable workers to track and analyse changes in land use and land cover patterns over a selected time period.

Sacred groves are community managed and therefore, human activity both within and in areas surrounding the groves could possibly have impacts of forest cover area over a selected time period, which can be observed using remotely sensed data. In this chapter, changes in forest cover of two sacred groves, and the change in land use land cover (LULC) of the surrounding areas are analysed using LANDSAT data for a period of 30 years from 1990 to 2020.

Remote Sensing and Forest Cover Change

Forests perform a number of important ecosystem services, serving as repositories for biodiversity, habitats for wildlife, carbon sinks, and climate regulation. They may, as in the case of sacred groves, be of social, cultural and/religious importance as well (Dorm-Adzobu *et al.*, 1991; Byers, Cunliffe and Huda, 2001; Ormsby, 2013; Diengdoh and Jayakumar, 2021). A change in forest cover can be indicative of corresponding change in efficacy with which forests perform important ecosystem services such as carbon sequestration and water resource management among others (Foley *et al.*, 2005). It may also, as in the case of sacred groves, be indicative of a change or shift in the socio-cultural milieu of the communities that manage them (Tiwari, Barik and Tripathi, 1998; Ormsby, 2013). The ability of satellite remote sensing to provide current and historical spatial data about a chosen study area at time interval selected by the researcher at their discretion is the reason behind its widespread use in various fields including that of forest cover mapping and change analysis (Adrien and Baumgardner, 1977).

A wide variety of research projects undertaken across a wide gamut of spheres have employed the use of satellite based remote sensing techniques including forest cover classification and mapping (Zhu and Evans, 1994; Laurin *et al.*, 2016; Yin *et al.*, 2017), forest cover quantification (Mayaux and Lambin, 1995; Tiwari, Sinha and Saran, 2016); community composition analysis (Martin *et al.*, 1998 ; Ferreira *et al.*, 2016); mapping the height of trees (Hansen *et al.*, 2016); forest cover change analysis and monitoring (Thiel *et al.*, 2006) and in the study of forest fires (Krishna and Reddy, 2012; Matin *et al.*, 2017). Remote sensing has also enabled workers to document and map and monitor areas that have been found to be vulnerable to forest fires (Chuvieco *et al.*, 2010; Ajin *et al.*, 2016).

Remote sensing has been used to monitor protected areas in India (Reddy, Jha and Dadhwal, 2013; Kaliraj, Meenakshi and Malar, 2012; Panigrahy *et al.*, 2010; Reddy, Jha and Dadhwal, 2016; Ranjan *et al.*, 2016; Bilyaminu *et al.*, 2021). The fact that spatial data is available for various time periods has made it possible to study ongoing occurrences such as forest degradation and deforestation (Jha, Dutt and Bawa, 2000; Da Ponte *et al.*, 2017). Remote sensing has also been used to understand the effects of activities such as the extraction of coal on vegetation over time (Sarma, 2005).

In the present study, geospatial data for a period of 30 years has been acquired to document the change in forest cover in two sacred groves, each of which is under the management of local community authority.

Remote Sensing and Land Use Land Cover Change

The availability of remote sensing data has also facilitated the study of how land is used by human beings, and how the human use of land can bring about changes in land cover. Causal relationships exist between LULC change and physical alterations to the environment and any changes in the provision of ecosystem services, which may be affected by and may in turn impact the socio-economic condition of an area (Butler, Corvalan and Koren, 2005). Past studies have elucidated possible causal relationships between LULC change and such problems as deforestation, loss of biodiversity (Mas *et al.*, 2004), considerably affecting the management of land as resource while also incurring various socio-economic impacts (Ojima, Galvin and Turner, 1994).

Spatiotemporal data have facilitated the analysis and monitoring of LULC change over extended periods of time, and enables researchers to study relationships between human activities, the LULC change they incur and impacts on the environment. Multiple studies in different countries have utilized remote sensing to monitor and track LULC change making extensive use of LANDSAT multi-temporal geospatial data (see Silapaswan *et al.*, 2001; Shalaby and Tateishi, 2007; Masek *et al.*, 2008; Dewan and Yamaguchi, 2009; Diallo, Hu and Wen, 2009; Thomas *et al.*, 2011; Hansen and Loveland, 2012; Mohajane *et al.*, 2018). Remote sensing techniques have also been employed to study the effect that LULC change can have on land surface temperatures (Jiang and Tian, 2010) and in the sphere of water resource management (Giardino *et al.*, 2010; Rai *et al.*, 2018) and have found application in the analysis and determination of the relationship between land use, land use change and the condition of water resources in an area (Butt *et al.*, 2015). Remote sensing has also been used to ascertain how the risk of soil erosion is influenced by land use change (Leh, Bajwa and Chaubey, 2013). Additionally, remote sensing techniques have been implemented to study LULC dynamics in areas affected by shifting cultivation (Vashum, 2016; Adhikary *et al.*, 2019).

Remote sensing is, therefore, a highly versatile tool that has been used in various spheres of research. As elucidated above, its application in LULC and forest cover change analysis indicate its suitability for studying the impact of human activities on forests and forest patches such as sacred groves.

In the present study, it was reasoned that human activities in the area surrounding the selected sacred groves could possibly have an impact on the aforementioned sacred groves. Ergo, any LULC change in the area surrounding each sacred grove could affect the grove concerned.

Therefore, in the present study, geospatial data collected over a period of 30 years allowed not only for the analysis of changes in forest cover in two selected sacred groves, but also for changes in LULC in a 1 km buffer region around the same in order to elucidate the interaction between LULC and the forest cover change dynamics of each grove.

Material and Method

Study Area

The two groves chosen for the study are situated in East Khasi Hills District, Meghalaya, found between 25°02″ and 26°07″N Lat. and 89°49″ and 92 °502″E Long. with a total area of 22,429 sq.km and at elevations ranging from approximately 50m to 1950m above sea level, constituting 0.7 per cent of the total area of the country.

East Khasi Hills District is found between 25°07″ and 25°41″N Latitude and 91°21″ and 92°09″E Longitude (Government of Meghalaya, 2022) with a total area of 2748 km^2 and is populated by a total of 8,24,059 people (Census, 2011). It comprises a total of 11 blocks and 1024 villages (Census, 2011) and is home to the state capital, Shillong.

For the purposes of the present study, two sacred groves and the immediate area surrounding them within a 1 km buffer area were selected. The groves selected were *Law Lyngdoh*, more commonly known as Mawphlang sacred grove, located at Mawphlang, and *'law Ryngkewlum Swer*, also known as Swer sacred grove, located at Swer village. The location of the study sites is shown in Figure 10.1.

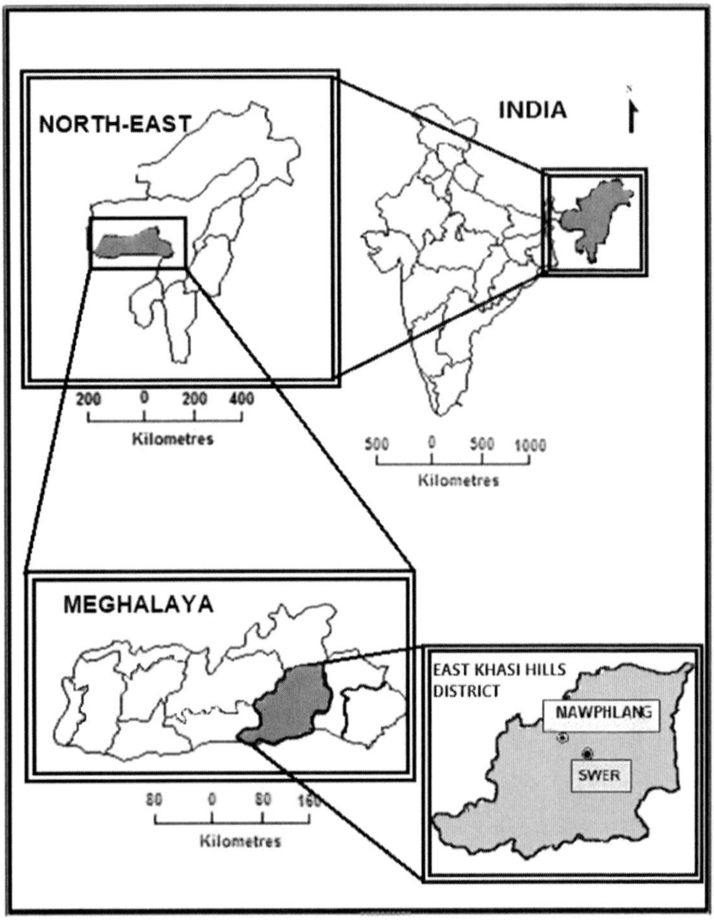

Fig. 10.1: Location Map of Study Area.

Law Lyngdoh Mawphlang, or Mawphlang sacred grove is located adjacent to Mawphlang village, which is approximately

25 km from the state capital, Shillong. The site is widely regarded as a well-known tourist destination and is marketed by the local community as an 'undisturbed' sacred forest. It is approximately of 78 ha in size and is considered to be a climax forest that is representative of the vegetation found occurring between 1500-2000m a.m.s.l. in North East India (Khan, Rai and Tripathi, 1986). Access to the premises is contingent upon the payment of an entry fee which may be inclusive of a guided tour of the grove itself. Activities such as unauthorized entry into the grove, and/or the collection of wood, non-timber forest products or any other article from the forest are strictly prohibited by local authorities such as the Mawphlang *Lyngdohship* (local indigenous priestly authority).

'Law Rynkewlum Swer or Swer sacred grove is located in close proximity to Swer village which is approximately 30 km south of Shillong (Mishra *et al.*, 2004) describes the grove as disturbed due to a variety of human activities taking including the grazing of livestock and the harvesting of wood. Located on a hill known as *lum Swer*, the grove is approximately of 59 ha in size and is surrounded by steep cliffs, patches of forest, grassland, cultivated land and settlement as well. Unlike Mawphlang sacred grove, Swer sacred grove has, as of the conducting of this study, not been marketed as a tourist destination, although Mawkdok bridge, a popular attraction is located in close proximity to the grove. Rules and regulations regarding access, entry and usage of the grove lack clarity and enforcement according to local authorities. As such, activities such as wood harvesting, though prohibited, and others such as the hunting and trapping of wildlife, though discouraged may take place.

Data Acquisition and Processing

In order to conduct the present study, the acquisition of multi-temporal data was required. Ergo, LANDSAT 5 and 8

Operational Land Imager (OLI) series satellite images were obtained for the time period 1990-2020 at intervals indicated below (see Table 10.1). The datasets were acquired from www.earthexplorer.usgs.gov. Use of data collected at relatively significant time intervals has been found to optimal for documenting change in forest cover (Tiwari *et al.*, 2013; Yismaw *et al.*, 2014; Ranjan *et al.*, 2016; Yadav *et al.*, 2022) and LULC (see Silapaswan *et al.*, 2001; Shalaby and Tateishi, 2007; Masek *et al.*, 2008; Dewan and Yamaguchi, 2009; Diallo, Hu and Wen, 2009; Thomas *et al.*, 2011; Hansen and Loveland, 2012; Yismaw *et al.*, 2014; Mohajane *et al.*, 2018) over a predetermined time period.

The use of LANDSAT datasets was preferred due to its medium resolution and the availability of geospatial time series data across multiple bands, which would facilitate and expedite LULC classification and change detection at the pixel level.

Data obtained *via* the following LANDSAT sensors were acquired: LANDSAT 5 and LANDSAT 8 OLI. Four cloud free images covering the area within which the selected study sites are included (Path-Row 136-042 and Path-row 137-042) were selected and downloaded from the LANDSAT archive data for the years 1990, 2000, 2014 and 2020. The details are tabulated below (see Table 10.1).

Table 10.1: Satellite data used for the study

Date of Acquisition	Sensor	Path/Row	Resolution
16-1-1990	LANDSAT 5	136/42	30m
12-1-2000	LANDSAT 5	136/42	30m
18-1-2014	LANDSAT 8 OLI	136/42	30m
30-3-2020	LANDSAT 8 OLI	137/42	30m

For LULC change analysis, it was decided that a buffer region of 1 km from each sacred grove would be delineated. The researchers' rationale was that the impact of human activities upon a sacred grove would increase with increasing proximity to said grove. Therefore, it was reasoned that activities occurring within a buffer area of 1 km would have a much greater impact on each respective sacred grove than those taking place at a greater distance away. Among these activities, crop cultivation, the construction of buildings and expansion of settlements, quarrying, harvesting of timber and/or firewood were included.

Analysis of changes in land use and land cover including agricultural land use, settlement, grassland, fallow land and forest cover necessitated the analysis of spatial data separated by considerable intervals of time to highlight important changes LULC in the immediate area around the study site. Therefore, the researchers selected a period of thirty years in order to elucidate and bring emphasis to, changes in forest cover within the groves and LULC in the surrounding area over a notably extended interval of time.

The software ERDAS IMAGINE 11 was used in the present study. A study area boundary around each site was delineated, and based on said boundary, a 1 km buffer boundary was also established around each grove. Accordingly, the boundary of each sacred grove was used to subset the images of Mawphlang and Swer sacred groves to facilitate the mapping of forest cover and forest cover change within each sacred grove. Similarly, the images obtained of the area surrounding each of the groves were subset with the buffer boundary as the outline for LULC classification within the buffer region. The images were then processed and false colour composites (FCCs) of the selected sacred groves and the surrounding buffer regions were created.

The flowchart showing the process of acquisition, preparation and analysis of data is given in Figure 10.2.

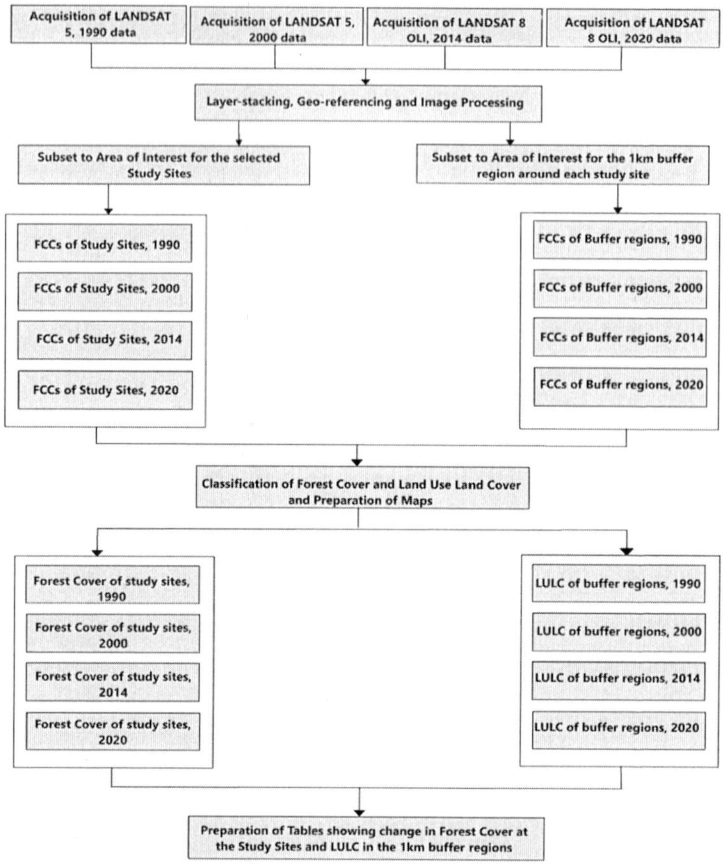

Fig. 10.2. Flow chart showing Data Acquisition, Processing, FCC Image Preparation, Map Preparation and Change Analysis in the present study.

For the present study, visual classification was selected as the method for change analysis instead of digital classification due to its demonstrated optimum efficiency and superior quality when used to carry out classification of medium-resolution data (Ghorbani and Pakravan, 2013).

Since the data obtained was of 30×30m resolution, classification was carried out at a suitable scale of 1:50,000.

However, for the purpose of clarity, the maps obtained have been enhanced to scales shown in the Figure 10.2. The FCCs obtained were then used to prepare maps documenting forest cover for the selected years at both sacred groves, and LULC within the 1 km buffer region of each site, for the same. Land Use and Land Cover in the buffer zone were grouped into following classes (see Table 10.2).

Table 10.2: Land Use and Land Cover classes in the 1 km buffer region around each site

No.	Classification	Comment/Description
1.	Community Forest	Area exclusively within the buffer zone that is under tree cover. Forests in the buffer region are managed by local communities and include village forests (*'law shnong*), community forests (*'law raij/'law raid*) and privately owned forests (*'law rikynti*) found in the area.
2.	Agricultural land	Land where cultivation of crops is taking place.
3.	Grassland	Land where grass is the dominant land cover.
4.	Settlement	Built-up land, characterized by the presence of man-man construction.
5.	Water bodies	Lotic ecosystems: rivers, streams; and Lentic ecosystems: lakes and reservoirs.
6.	Others	Stone/sand quarries and other land that cannot be classified.

Note: Accuracy for the mapping of forest cover was 88 per cent while the accuracy for LULC classification was 85 per cent.

Results and Discussion

Forest Cover Classification and Change Detection within the Selected Sites

Mawphlang sacred grove was observed to primarily show the presence of dense forest, with significantly smaller patches of open forest. In the case of Swer sacred grove, however, while the largest area observed was classified under dense forest, significant area was also classified as open and degraded forest. FCC images of the two sites are shown in Figure 10.3, and forest cover classification images of the two sacred groves in Figure 10.4.

No degraded forest was observed at Mawphlang for all datasets, and it was notable that the change from dense forest to open forest (which may be interpreted as decline) during the entirety of the selected study period was less than 1 per cent.

At Swer, however, while minimal change in degraded forest was observed during the first two time intervals (1990-2000 and 2000-2014), a comparatively considerable increase in degraded forest was observed between 2014 and 2020. A steady decline in open forest was observed between the years 1990 and 2000. A slight increase in open forest (0.77%) during the second time interval (2000-2014) following a decline from 1990-2000 (0.76%) may be attributed to pixel error, and can be overlooked. A decline (0.27%) was observed during the third time interval (2014-2020) when compared to the second interval and corresponds the aforementioned increase in area under degraded forest.

The area classified as dense forest was observed to be in steady decline throughout the entirety of selected study period with the exception of a minor increase (0.06%) from 2000-2014 which may likely be caused by pixel error or due to the implementation of tree plantation programmes, which were then offset by a comparatively greater (5.81%) loss in forest cover during the third time period (2014-2020). During the third time interval, area under dense forest experienced a loss of 3.4 ha which is

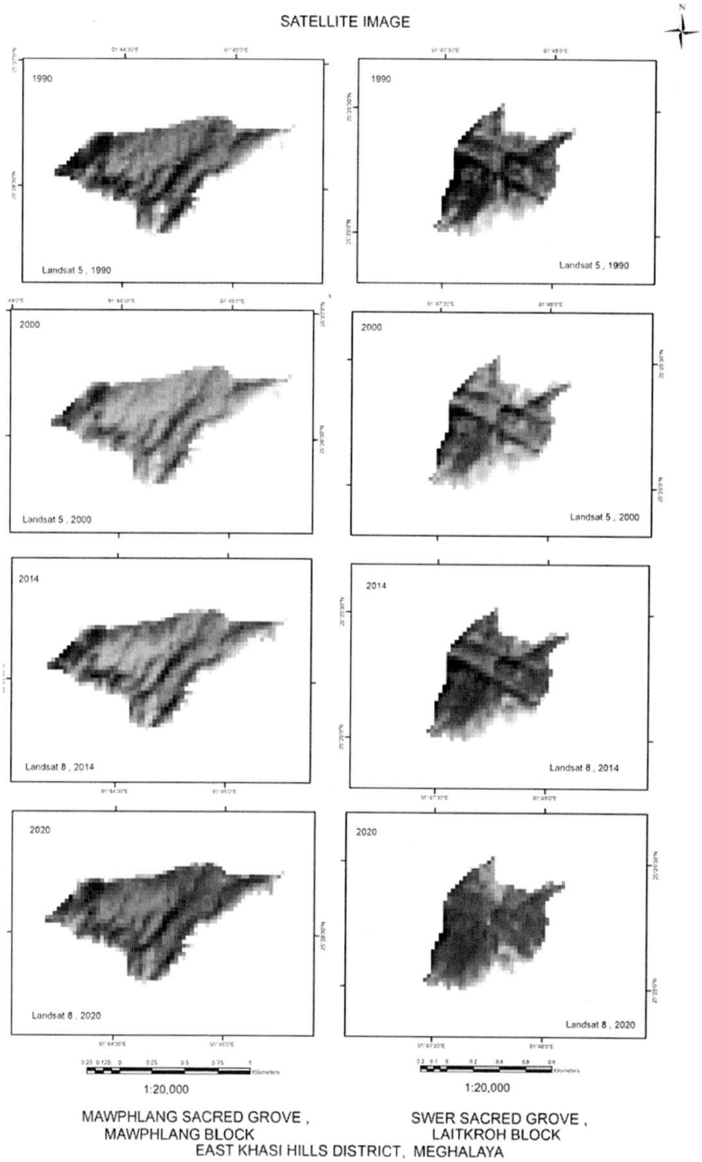

Fig. 10.3: False Colour Composites of the study areas from 1990 to 2020.

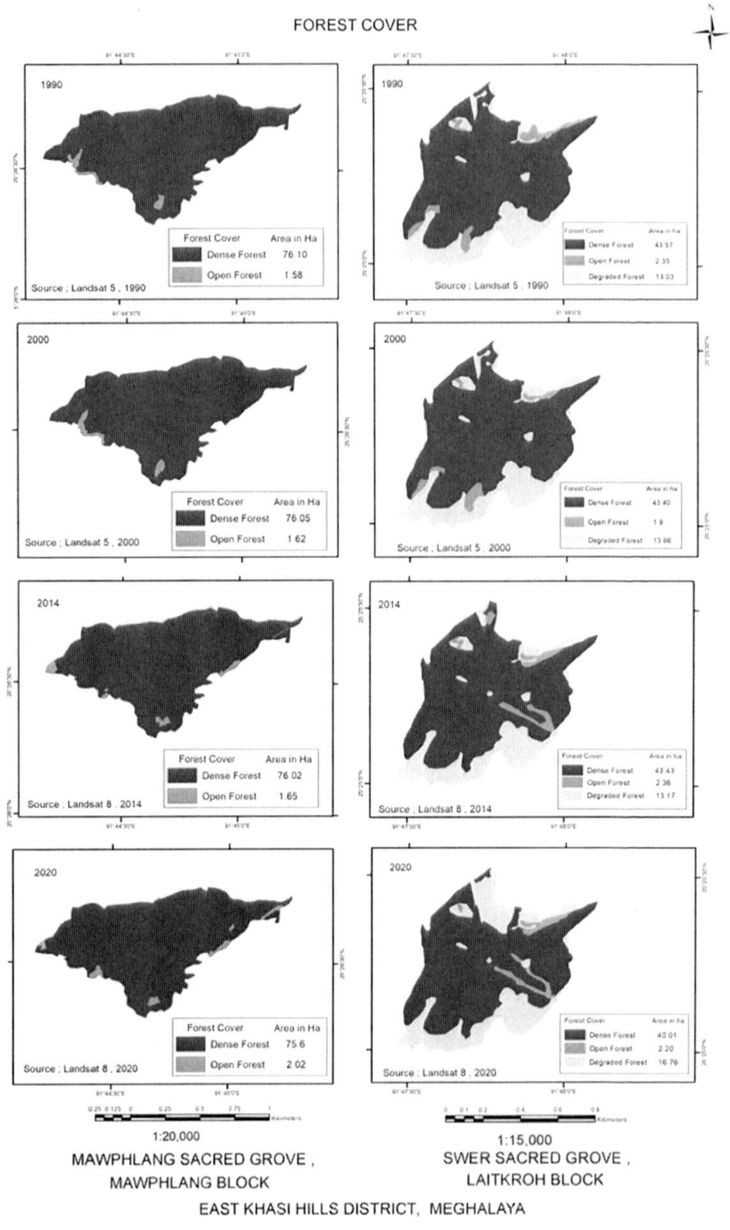

Fig. 10.4. Forest Cover of the study areas from 1990 to 2020.

equal to 7.8 per cent of the total area under dense forest in the year 1990.

The present observation of degraded forest corroborates the findings of an earlier study (Mishra *et al.*, 2004) which had classified different areas of the grove based on levels of disturbance and classified some areas of the grove as under high disturbance. Forest cover classification and change in forest cover for Mawphlang and Swer sacred groves over the selected study period are shown in Table 10.3 and Table 10.4 respectively. Percentage changes in forest cover at Mawphlang and Swer over the course of the study period taking 1990 as the baseline are shown in Table 10.3.

Table 10.3: Forest Cover of Mawphlang sacred grove and Swer sacred grove (Area in Ha)

Forest Cover Class	1990		2000		2014		2020	
	Maw-phlang	Swer	Maw-phlang	Swer	Maw-phlang	Swer	Maw-phlang	Swer
Dense Forest	76.1	43.6	76.1	43.4	76	43.4	75.7	40
Open Forest	1.6	2.4	1.6	1.9	1.7	2.4	2	2.2
Degraded Forest	–	13	–	13.7	–	13.2	–	16.8

Table 10.4: Change in Forest cover of Mawphlang sacred grove and Swer sacred grove from 1990 to 2020 (%)

Forest Cover Class	1990-2000		2000-2014		2014-2020	
	Mawphlang	Swer	Mawphlang	Swer	Mawphlang	Swer
Dense Forest	-0.06	-0.3	-0.03	0.06	-0.48	-5.81
Open Forest	0.06	-0.76	0.03	0.77	0.48	-0.27
Degraded Forest	–	1.06	–	-0.83	–	6.08

Table 10.5: Change in Forest cover of Mawphlang sacred grove and Swer sacred grove from 1990 to 2020 (% change from 1990 baseline)

Forest Cover Class	1990-2000		1990-2014		1990-2020	
	Mawphlang	**Swer**	**Mawphlang**	**Swer**	**Mawphlang**	**Swer**
Dense Forest	-0.06	-0.3	-0.09	-0.24	-0.57	-6.05
Open Forest	0.06	-0.76	0.09	0.01	0.57	-0.26
Degraded Forest	–	1.06	–	0.23	–	6.31

The extraction or harvesting of any resources from a sacred grove is usually restricted (Tiwari *et al.*, 2013). However, the enforcement of this stipulation largely depends upon the practices of the concerned village authorities. Both communities deny extracting wood from their sacred groves (Diengdoh, 2022).

At Mawphlang, strict regulations prohibit the harvesting or collection of wood or any Non-Timber Forest Products (NTFPs) from within the sacred grove. Edible NTFPs, however, may be consumed but only within the grove. Change in forest cover in the grove from dense to open may be due to natural causes such as weather events or undetected intrusion and harvesting. It may not be feasible for communities to provide round the clock monitoring of the grove.

'Law Ryngkew lum Swer, however, is not a tourist attraction. Rules and regulations lack clarity, have, as of the conducting of this study, not been notarized, or codified, and have not been observably enforced. Entry into the grove is neither monitored nor restricted, indicating that the implementation and protection regimes of sacred groves is contingent upon the village authorities, and shows disagreement with the findings of Tiwari *et al.* (2013). Notable tracts of the grove show degradation. The

unsanctioned harvesting of wood by intruders and the impacts of natural events such as storms and lightning strikes cannot be discounted. The borders of the grove show signs of encroachment for the purpose of house construction by local community members. The northern edge of the grove shows signs of recent degradation (2014-2020), most likely due to human activities. While local communities have denied harvesting wood from the grove (Diengdoh, 2022), the possibility of such activities taking place without the permission or knowledge of the concerned authority, possibly carried out by intruders from other nearby settlements cannot be ruled out. Loss of area under dense forest cover and increased area under open forest has been observed in studies in other parts of the country, at varying scales (Jha, Dutt and Bawa, 2000; Bilyaminu *et al.*, 2021). A steady decline in very dense forest was also observed by Ranjan *et al.* (2016) from 2009-2016 in Dalma Wildlife Sanctuary with possible causal factors including harvesting of wood, which though denied by respondents at Swer (Diengdoh, 2022) has been discussed in past studies conducted at the site (see Mishra *et al.*, 2004). Past studies in the state (Tiwari, Barik and Tripathi, 1998; Ormsby 2013) indicate that religious and socio-cultural changes have resulted in changes in how groves are viewed and, by extension how they are managed. A number of sacred groves in Meghalaya have suffered great degradation, which has been linked to a religious and cultural shift, from the indigenous belief system to Christianity (Ormsby, 2013).

At Mawphlang, however, the grove has remained largely intact, which corroborates the observations of Tiwari, Barik and Tripathi (1998), who describe it as an "undisturbed" (p. 23) site. Findings indicate that the grove has largely remained unaffected by human activities being carried out in the buffer region around it. It is highly probable that the strict enforcement of existing regulations and the belief of local communities in the existence of a sylvan deity or spirit (*u ryngkewu basa*), who may punish violators have played a strong role to maintain the integrity of the grove (Diengdoh and Jayakumar, 2021; Diengdoh, 2022). The fact that the site is marketed as a popular tourist destination by local communities who have come to regard sacred groves as possible sources of income and employment may also have an important role to play in lending impetus to conservation efforts and protection regimes (Diengdoh and Jayakumar, 2021; Diengdoh, 2022). Efforts by authorities to improve protection regimes have been found to contribute to the stability of forest cover (Mohajane *et al.*, 2018), and the observations at Mawphlang sacred grove may reflect this.

The institution of a protected area can help to preserve and even bring about an increment in area under forest cover (Fox and Vogler, 2005). It must be noted, however, that the efficacy of such declarations in the protection of forests depend upon how effectively rules and regulations are enforced. Activities such as shifting cultivation, encroachment of human settlements, and occurrences that may be of human or natural origin such as forest fires have been found to cause forest degradation in protected areas (Ranjan *et al.*, 2016), especially, if unrestrained by strongly enforced regulations. The expansion of settlements and the pressure to increase land area under cultivation have been discussed as probable causes for considerable forest degradation (Bilyaminu *et al.*, 2021). It has also been noted that a simultaneous loss in dense forest coinciding with an increase in area under open forest and scrub land can be inferred to be indicators of pressure being incurred on a forest (Panigrahy *et al.*, 2010). It is highly probable that the clarity of the regulations governing the use of the sacred grove at Mawphlang, and the strictness with which they have been implemented and adhered to constitute an important factor contributing to the grove remaining comparatively intact with minimal disturbance.

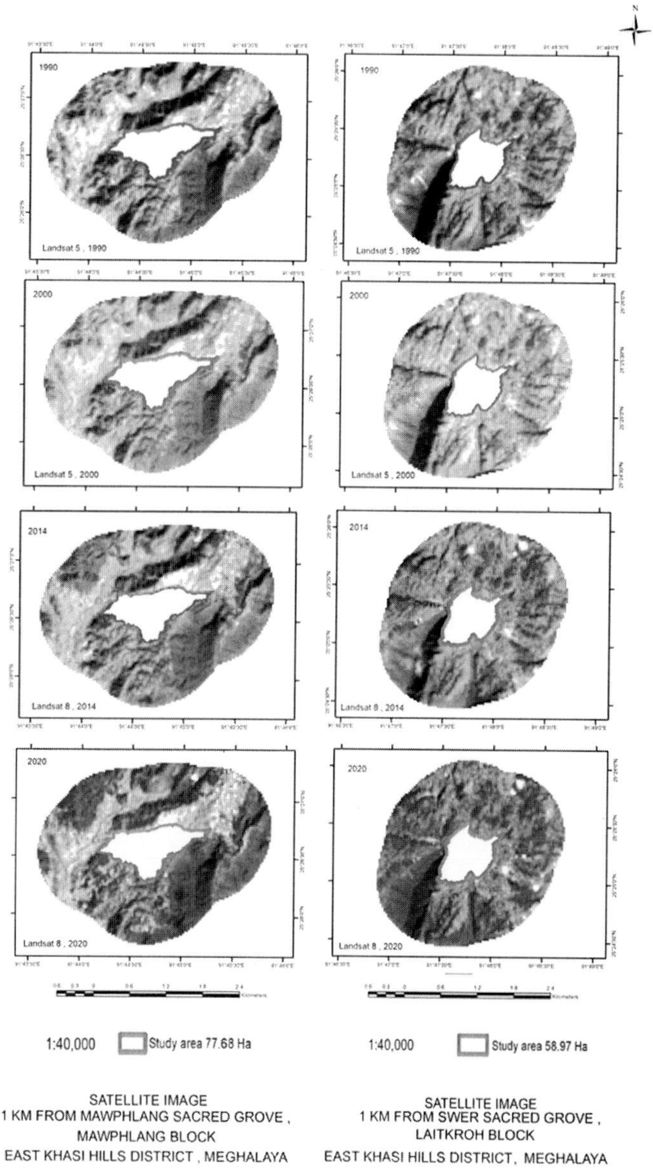

Fig. 10.5: False Colour Composite images of 1 km buffer region of both sites from 1990 to 2020.

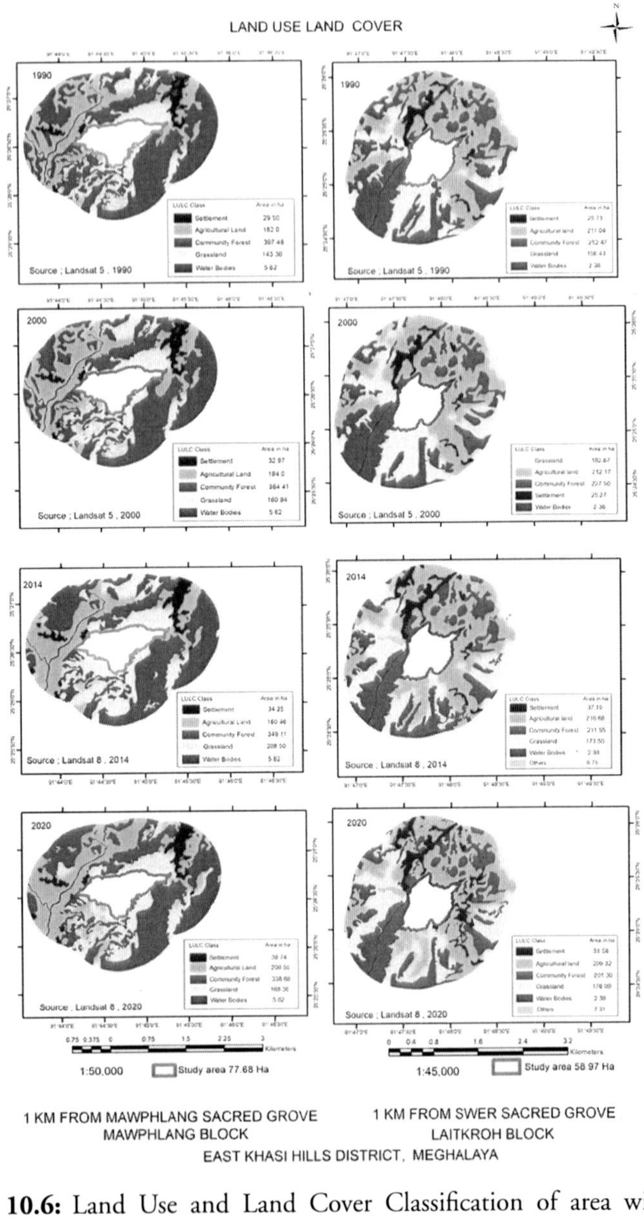

Fig. 10.6: Land Use and Land Cover Classification of area within 1 km buffer region around Swer sacred grove from 1990 to 2020.

At Swer, however, the lack of clear, enforceable rules may have lent impetus to the degradation of the grove by facilitating activities such as intrusion, wood harvesting and hunting.

Land Use Change in the Buffer Region

The dynamics of the socio-economic milieu of a community are often reflected in how LULC patterns in the area occupied by said community change over time (Diallo, Hu and Wen, 2009). LULC patterns in the 1 km buffer region around both sites showed notable changes over the course of the selected study period. The FCCs for the 1 km buffer area around both sites are shown in Figure 11.5 while the corresponding LULC maps are shown in Figure 9.6.

The area under each LULC class within the 1 km buffer regions around both sites for the selected study period is shown in Table 10.6, and changes in area under the various classes for the both areas are shown in Table 10.7. Percentage changes in LULC over the course of the study period taking 1990 as a baseline for Mawphlang and Swer are shown in Table 10.8.

Table 10.6: Land Use and Land Cover classification within 1 km buffer region at Mawphlang and at Swer (Area in Ha)

LULC Class	1990		2000		2014		2020	
	Maw-phlang	Swer	Maw-phlang	Swer	Maw-phlang	Swer	Maw-phlang	Swer
Settlement	29.5	25.73	33	25.56	34.3	37.16	38.7	51.59
Agricultural land	182	211.04	194	212.18	160.5	218.66	206.6	209.32
Community Forest	397.5	252.48	364.4	227.5	349.1	211.56	338.7	201.31
Grassland	143.4	158.38	160.9	182.39	208.5	173.5	168.4	178.1
Waterbodies	5.6	2.38	5.6	2.38	5.6	2.38	5.6	2.38
Others	–	0	–	0	–	6.76	-	7.31

Table 10.7: Change in Land Use and Land Cover within 1 km buffer region at Mawphlang between 1990 and 2020 (Change in %)

LULC Class	1990-2000		2000-2014		2014-2020	
	Maw-phlang	**Swer**	**Maw-phlang**	**Swer**	**Maw-phlang**	**Swer**
Settlement	0.46	-0.03	0.17	1.79	0.59	2.22
Agricultural land	1.59	0.17	-4.43	1	6.08	-1.44
Community Forest	-4.36	-3.84	-2.02	-2.45	-1.38	-1.58
Grassland	2.32	3.69	6.28	-1.37	-5.3	0.71
Waterbodies	0	0	0	0	0	0
Others	–	0	–	1.04	–	0.09

Table 10.8: Change in Land Use and Land Cover Classification within 1 km buffer region at Mawphlang and at Swer (Change in % from 1990 baseline)

LULC Class	1990-2000		1990-2014		1990-2020	
	Maw-phlang	**Swer**	**Maw-phlang**	**Swer**	**Maw-phlang**	**Swer**
Settlement	0.46	-0.03	0.63	1.76	1.22	3.98
Agricultural land	1.59	0.17	-2.84	1.17	3.24	-0.27
Community Forest	-4.36	-3.84	-6.38	-6.29	-7.76	-7.87
Grassland	2.32	3.69	8.6	2.32	3.3	3.03
Waterbodies	0	0	0	0	0	0
Others	–	0	–	1.04	–	1.13

At Mawphlang, it was observed that an increase in area under grassland would coincide with a decline in area under cultivated/agricultural land and *vice versa*. Area under community forests, however, showed a steady decline throughout the study period. A decline in community forest may correspond to an increase in area under agriculture or grassland, as conversion of forested land to farmland during the study period cannot be ruled out. However, any inverse relationship between a change in area under cultivation, and area that is classified as grassland may be due to land leasing that takes place in the area under which system, land, such as grassland, may be leased to local families or individuals for cultivation by private owners or community authorities for a fixed period of time. Should the lease elapse and not be renewed the cultivated land may be left fallow and becomes grassland again. Such a possibility was highlighted by Lyngkhoi *et al.* (2022) in the state regarding land under broom grass cultivation reverting to community administration and being left unused after cultivation. However, the dynamics of land leasing in the area, however, do not fall within the scope of the present study and may be the subject of future research in the area.

At Swer, however, this pattern was not observed. Agricultural land demonstrated minimal change while settlements were observed to have increased during the period of 2000-2014 and 2014-2020 respectively. An apparent decrease of 0.03 per cent in area under settlement during the first ten-year interval (1990-2000) may be attributed to pixel error and has been disregarded. Area under water bodies showed no change during the entire study period. It is possible that cultivation has remained restricted to areas either privately owned or designated for cultivation by the community. The decrease in area under agriculture during the third time interval (2014-2020) may be caused by land being left fallow following consecutive seasons of cultivation, or by the expansion of settlements or by tree plantation activities during the same.

At Swer, the area under settlements increased notably during the second and third time periods (2000-2014 and 2014-2020), while at Mawphlang, only comparatively minor increments were observed. Water bodies in both areas demonstrated no observable change. Area under others, encompassing quarries and land then cannot be identified, was found only in the buffer region around Swer.

At both sites, area under community forest was found to decrease steadily throughout the 30-year period in keeping with other studies in the country (Panigrahy *et al.*, 2010; Kale *et al.*, 2016; Reddy, Jha and Dadhwal, 2016; Ranjan *et al.*, 2016; Bilyaminu *et al.*, 2021) and the region (Fox and Vogler, 2005; Diallo, Hu and Wen, 2009) although the work of Kale *et al.* (2016) indicates that the rate of loss is slowing possibly due to conservation efforts (Reddy, Jha and Dadhwal, 2016). The most probable cause for decline is the harvesting of timber and/or wood may fall under the "basic needs" (Ranjan *et al.*, 2016), which communities may venture into forests to fulfil. The conversion of forested land to cultivated land has been documented in other studies (Panigrahy *et al.*, 2010; Yismaw *et al.*, 2014; Ranjan *et al.*, 2016) and is among the possible causes for decline of forest cover in protected areas as well (Bilyaminu *et al.*, 2021).

At Mawphlang, notably, the decline continued from 2014-2020 in spite of the institution of the Khasi Hills REDD+ project (see Poffenberger, 2014, 2015) which has been undertaken by local community's authorities in the area with the aim of regulating and reducing the harvesting of wood and shifting cultivation in the area. The effectiveness of REDD+ has been shown to vary considerably, with some programmes demonstrating limited success in reducing forest cover loss especially if revenue from carbon sequestration has not materialized (Collins *et al.*, 2022) or if pledges of payment from interest groups fail to meet the expectations of local stakeholders (Fletcher *et al.*, 2016). Another possibility for the decline may be due to a less than optimum efficacy in efforts to prevent forest fires and other calamities of natural or human origin.

At Swer, community forest was observed to have declined notably during the second time interval (2000-2014), likely due to considerable wood harvesting, and land clearing for agriculture. The threat posed to the grove by resource harvesting and other human activities was highlighted by a previous study (Mishra *et al.*, 2004). The effectiveness of community based forest management regimes in inculcating in local communities the importance of using forest resources sustainably has been found to be highly limited (Nongkynrih, 2006). Therefore, protocols regarding restoration of forest resources either lack clarity, have been left to the discretion of communities or private owners, or disregarded altogether (Diengdoh and Wahlang, 2016).

Community managed forests such as those found within the buffer regions as delineated in the present study are classified and designated based on what they are used for and how by the communities that manage them (District Council Forest Act, 1958; Dasgupta and Symlieh, 2006). For example, access to '*law shnong* (village forests) at Mawphlang for the purpose of extraction of NTFPs or a fixed quantity of timber is granted to permanent members of the community, while the usufruct rights of visitors or temporary residents or visitors may be restricted only to certain NTFPs and water resources (Nongkynrih, 2006). Furthermore, community forest types such as *'law ri Kynti* (privately owned forests) are each owned and managed by a particular family, clan or a number of joint clans (Tiwari, Tynsong and Lynser, 2010). Usage of such forests would thus be determined by the owners/managing family/clan or joint clan, at their discretion. Protection regimes in forests such as '*law ri Kynti, 'law shnong* or *'law raid* among others have been found to be lacking (Tiwari *et al.*, 2013). This may be a contributing factor to the loss of forest cover in the buffer region at both sites.

Furthermore, the irregularity of shape, increased patchiness and overall fragmented distribution of community forests in the buffer region around Swer sacred grove may be contributing to edge effect in the area, and may also have a negative impact on species as has been found by other workers (Ewers and Didham, 2006) with edge effect in particular adversely affecting various ecological processes and hindering the ecosystem services (de Paula, Groeneveld and Huth, 2016).

Conclusion

Analysis of geospatial data in the present study indicates a notable disparity between the two selected study sites. More extensive forest cover change, and degradation was observed at Swer than at Mawphlang, which in turn may be indicative of the important role that effectively enforced community-based protection regimes play in preserving sacred groves. Furthermore, it may be inferred that said effectiveness is contingent, to varying extents, upon the enforcement of such regimes not only within the boundary of each grove but also upon the management regimes of the surrounding lands, and how efficaciously they are upheld. Such regimes are further subject to a number of other local factors such as rules and regulations governing *inter alia* land tenure, land ownership and leasing. It was observed that Mawphlang sacred grove has remained comparatively well preserved during the entirety of the study period while Swer sacred grove has shown notable decline, as reflected by the decrease in area under dense forest and a corresponding increase in area under degraded and open forest during the same period. This decline may be attributed to a lack of clarity and enforcement of protection regimes resulting in damage being observed at the grove, despite local communities denying that they exploit their grove for timber resources (Diengdoh and Jayakumar, 2021; Diengdoh, 2022). Intrusion and extraction by outsiders from neighbouring villages is also possible. The observed decline in community forest area within both buffer regions may be caused by a number of factors such as a lack of protection regimes governing the use of community forests, an ineffectual implementation of the same, or that the extraction of resources from such forests has not been offset by such compensatory measures as tree plantation or, *inter alia*, the assisted natural regeneration of degraded forested areas. A conversion of forest land to agricultural land or housing land cannot be ruled out. Land leasing practices may also constitute in an important causal factor for the notable increase in agricultural land area and coinciding decrease in grassland area, and *vice versa*.

Overall, the present study indicates that the conservation and management of sacred groves for the present and for posterity is subject to the influence of human activities. It also demonstrates the applicability of remote sensing as a versatile tool for monitoring, analysing and better understanding the interaction between said human activities and the changes they may incur in the surrounding environment.

REFERENCES

Adhikary, Partha Pratim, D. Barman, M. Madhu, Ch J. Dash, P. Jakhar, H.C. Hombegowda, B.S. Naik, D.C. Sahoo, and Karma Beer (2019). "Land Use and Land Cover Dynamics with Special Emphasis on Shifting Cultivation in Eastern Ghats Highlands of India Using Remote Sensing Data and GIS." *Environmental Monitoring and Assessment, 191*: 1-15.

Adrien, Pierre-Marie, and Marion F. Baumgardner (1977). "Landsat, Computers, and Development Projects: Digital Analysis of Data from Earth-Orbiting Satellites Offers New Perspectives on Global Development." *Science, 198* (4316): 466-470.

Ajin, R.S., Ana-Maria Loghin, Mathew K. Jacob, P.G. Vinod, and R.R. Krishnamurthy (2016). "The Risk Assessment Study of Potential Forest Fire in Idukki Wildlife Sanctuary Using RS and GIS

Techniques". *International Journal of Advanced Earth Science and Engineering, 5*(1): 308-318.

Bilyaminu, H., P. Radhakrishnan, K. Vidyasagaran, and K. Srinivasan. (2021). "Monitoring Land Use and Land Cover Change of Forest Ecosystems of Shendurney Wildlife Sanctuary, Western Ghats, India". *Asian Journal of Environment and Ecology, 15*(4): 20-27.

Butler, Colin, D., Carlos F. Corvalan and Hillel S. Koren (2005). "Human Health, Well-being, and Global Ecological Scenarios." *Ecosystems, 8*: 153-162.

Butt, Amna, Rabia Shabbir, Sheikh Saeed Ahmad and Neelam Aziz (2015). "Land Use Change Mapping and Analysis Using Remote Sensing and GIS: A Case Study of Simly Watershed, Islamabad, Pakistan." *The Egyptian Journal of Remote Sensing and Space Science 18*(2): 251-259.

Byers, Bruce A., Robert N. Cunliffe and Andrew T. Hudak (2001). "Linking the Conservation of Culture and Nature: A Case Study of Sacred Forests in Zimbabwe." *Human Ecology, 29*: 187-218.

Chandramouli, C. and Registrar General (2011). "Census of India." *Rural Urban Distribution of Population, Provisional Population Total.* New Delhi: Office of the Registrar General and Census Commissioner, India.

Chuvieco, Emilio, Inmaculada Aguado, Marta Yebra, Héctor Nieto, Javier Salas, M. Pilar Martín, Lara Vilar *et al.* (2010). "Development of a Framework for Fire Risk Assessment Using Remote Sensing and Geographic Information System Technologies." *Ecological Modelling, 221*(1): 46-58.

Collins, Amy C., Mark N. Grote, Tim Caro, Aniruddha Ghosh, James Thorne, Jonathan Salerno and Monique Borgerhoff Mulder (2022). "How Community Forest Management Performs when REDD+ Payments Fail". *Environmental Research Letters, 17*(3): 034019.

Da Ponte, Emmanuel, Benjamin Mack, Christian Wohlfart, Oscar Rodas, Martina Fleckenstein, Natascha Oppelt, Stefan Dech and Claudia Kuenzer (2017). "Assessing Forest Cover Dynamics and Forest Perception in the Atlantic Forest of Paraguay, Combining Remote Sensing and Household Level Data." *Forests, 8*(10): 389.

Dasgupta, J. and Symlieh, H.J. (2006). *Trends in Tenure Arrangements for Forest and their Implications for Sustainable Forest Management—The Need for a More Unified Regime: A Case Study from Meghalaya, India.* Rome, FAO.

Dewan, Ashraf M. and Yasushi Yamaguchi (2009)."Land Use and Land Cover Change in Greater Dhaka, Bangladesh: Using Remote Sensing to Promote Sustainable Urbanization." *Applied Geography, 29*(3): 390-401.

de Paula, Mateus Dantas, Jürgen Groeneveld and Andreas Huth (2016). "The Extent of Edge Effects in Fragmented Landscapes: Insights from Satellite Measurements of Tree Cover." *Ecological Indicators, 69*: 196-204.

Diallo, Yacouba, Guangdao Hu and Xingping Wen (2009). "Applications of Remote Sensing in Land Use/Land Cover Change Detection in Puer and Simao Counties, Yunnan Province." *Journal of American Science, 5*(4): 157-166.

Diengdoh, Bennathaniel Hynniewta and S. Jayakumar (2021). "A Comparative Study of Local Attitudes and Perceptions of Sacred Groves between Village Communities in East Khasi Hills, Meghalaya, India." *International Journal of Ecology and Environmental Sciences, 47*(3): 197-208.

Diengdoh, Bennathaniel H. and Banphira Bahun Wahlang (2016). "Climate Change Migration and Traditional Institutions: Can REDD+ Revitalise Age-old Khasi Regimes of Community Based Forest Management?" *Revisiting Traditional Institutions in the Khasi-Jaintia Hills*, pp. 200-226.

Diengdoh, Bennathaniel H. (2016). "A Comparative Study of Biodiversity and Human Interactions in two Sacred Groves in East Khasi Hills District, Meghalaya, India." Ph.D. dissertation, Department of Ecology and Environmental Sciences, Pondicherry University, Puducherry.

District Council Forest Act (1958). "The United Khasi-Jaintia Hills Autonomous District (Management and Control of Forests) Act, 1958 (Act 1 of 1958)." *Assam Gazette*, January 22.

Dorm-Adzobu, Clement, Okyeame Ampadu-Agyei, and Peter G. Veit. (1991). *Religious Beliefs and Environmental Protection: The Malshegu Sacred Grove in Northern Ghana.* No. 4. Acts Press.

Ewers, Robert M. and Raphael K. Didham (2006)."Confounding Factors in the Detection of Species Responses to Habitat Fragmentation." *Biological Reviews, 81*(1): 117-142.

Ferreira, Matheus Pinheiro, Maciel Zortea, Daniel Capella Zanotta, Yosio Edemir Shimabukuro and Carlos Roberto deSouza Filho.

(2016). "Mapping Tree Species in Tropical Seasonal Semi-deciduous Forests with Hyperspectral and Multispectral Data." *Remote Sensing of Environment, 179*: 66-78.

Fletcher, Robert, Wolfram Dressler, Bram Büscher and Zachary R. Anderson (2016). "Questioning REDD+ and the Future of Market-based Conservation." *Conservation Biology, 30*(3): 673-675.

Foley, Jonathan A., Ruth DeFries, Gregory P. Asner, Carol Barford, Gordon Bonan, Stephen R. Carpenter, F. Stuart Chapin *et al.* (2005). "Global Consequences of Land Use." *Science, 309*(5734): 570-574.

Fox, Jefferson and John B. Vogler (2005). "Land-use and Land-cover Change in Montane Mainland Southeast Asia." *Environmental Management, 36*: 394-403.

Ghorbani, Ardavan and Mina Pakravan (2013). "Land Use Mapping Using Visual *vs.* Digital Image Interpretation of TM and Google Earth Derived Imagery in Shrivan-Darasi Watershed (North-west of Iran)." *European Journal of Experimental Biology, 3*(1): 576-582.

Giardino, Claudia, Mariano Bresciani, Paolo Villa and Angiolo Martinelli (2010). "Application of Remote Sensing in Water Resource Management: The Case Study of Lake Trasimeno, Italy." *Water Resources Management, 24*: 388-389.

Government of India (2022). *East Khasi Hills*. https://eastkhasihills.gov.in/

Hansen, Matthew C. and Thomas R. Loveland (2012). "A Review of Large Area Monitoring of Land Cover Change Using Landsat Data." *Remote Sensing of Environment, 122*: 66-74.

Hansen, Matthew C., Peter V. Potapov, Scott J. Goetz, Svetlana Turubanova, Alexandra Tyukavina, Alexander Krylov, Anil Kommareddy and Alexey Egorov (2016). "Mapping Tree Height Distributions in Sub-Saharan Africa Using Landsat 7 and 8 Data." *Remote Sensing of Environment, 185*: 221-232.

Jha, C.S., C.B.S. Dutt and K.S. Bawa (2000). "Deforestation and Land Use changes in Western Ghats, India." *Current Science*, pp. 231-238.

Jiang, Jing and Guangjin Tian (2010). "Analysis of the Impact of Land Use/Land Cover Change on Land Surface Temperature with Remote Sensing." *Procedia Environmental Sciences, 2*: 571-575.

Kale, Manish P., Manoj Chavan, Satish Pardeshi, Chitiz Joshi, Prabhakar A. Verma, P.S. Roy, S.K. Srivastav *et al.* (2016). "Land-use and Land-cover Change in Western Ghats of India." *Environmental Monitoring and Assessment, 188*: 1-23.

Kaliraj, S., S. Muthu Meenakshi and V.K. Malar (2012). "Application of Remote Sensing in Detection of Forest Cover Changes Using Geo-Statistical Change Detection Matrices—A Case Study of Devanampatti Reserve Forest, Tamil Nadu, India." *Nature, Environment and Pollution Technology, 11*(2): 261-269.

Khan, M.L., Ashalata Devi Khumbongmayum and R.S. Tripathi (2008). "The Sacred Groves and their Significance in Conserving Biodiversity: An Overview." *International Journal of Ecology and Environmental Sciences, 34*(3): 277-291.

Khan, M.L., J.P.N. Rai and R.S. Tripathi (1986). "Regeneration and Survival of Tree Seedlings and Sprouts in Tropical Deciduous and Sub-Tropical Forests of Meghalaya, India." *Forest Ecology and Management 14*(4): 293-304.

Krishna, P. Hari and C. Sudhakar Reddy (2012). "Assessment of Increasing Threat of Forest Fires in Rajasthan, India Using Multi-Temporal Remote Sensing Data." *Current Science*, pp. 1288-1297.

Laurin, Gaia Vaglio, Nicola Puletti, William Hawthorne, Veraldo Liesenberg, Piermaria Corona, Dario Papale, Qi Chen and Riccardo Valentini (2016). "Discrimination of Tropical Forest Types, Dominant Species and Mapping of Functional Guilds by Hyperspectral and Simulated Multispectral Sentinel-2 Data." *Remote Sensing of Environment, 176*: 163-176.

Leh, M., S. Bajwa and I. Chaubey (2013). "Impact of Land Use Change on Erosion Risk: An Integrated Remote Sensing, Geographic Information System and Modeling Methodology." *Land Degradation & Development, 24*(5): 409-421.

Lyngkhoi, Dipriya R., S. Basanta Singh, Ram Singh and Bhanita Baruah (2022)."Shifting Cultivation to Settled Agriculture: Land Ownership Rights and Cropping Pattern in Meghalaya, North East India." *Indian Res. J. Ext. Edu. 22*(5): 150-154.

Martin, M.E., Steven D. Newman, John D. Aber and Russell G. Congalton (1998). "Determining Forest Species Composition Using High Spectral Resolution Remote Sensing Data." *Remote Sensing of Environment, 65*(3): 249-254.

Mas, Jean-Francois, Alejandro Velázquez, José Reyes Díaz-Gallegos, Rafael Mayorga-Saucedo, Camilo Alcántara, Gerardo Bocco, Rutilio Castro, Tania Fernández and Azucena Pérez-Vega (2004). "Assessing Land Use/Cover Changes: A Nationwide Multidate Spatial Database for Mexico." *International Journal of Applied Earth Observation and Geoinformation, 5*(4): 249-261.

Masek, Jeffrey G., Chengquan Huang, Robert Wolfe, Warren Cohen, Forrest Hall, Jonathan Kutler and Peder Nelson (2008). "North American Forest Disturbance Mapped from a Decadal Landsat Record." *Remote Sensing of Environment, 112*(6): 2914-2926.

Matin, Mir A., Vishwas Sudhir Chitale, Manchiraju S.R. Murthy, Kabir Uddin, Birendra Bajracharya and Sudip Pradhan. (2017). "Understanding Forest Fire Patterns and Risk in Nepal Using Remote Sensing, Geographic Information System and Historical Fire Data." *International Journal of Wildland Fire, 26*(4): 276-286.

Mayaux, Philippe and Eric F. Lambin (1995)."Estimation of Tropical Forest Area from Coarse Spatial Resolution Data: A Two-step Correction Function for Proportional Errors due to Spatial Aggregation." *Remote Sensing of Environment, 53*(1): 1-15.

Mishra, B.P., O.P. Tripathi, R.S. Tripathi and H.N. Pandey (2004). "Effects of Anthropogenic Disturbance on Plant Diversity and Community Structure of a Sacred Grove in Meghalaya, North East India." *Biodiversity & Conservation, 13*: 421-436.

Mohajane, Meriame, A.L.I. Essahlaoui, Fatiha Oudija, Mohammed El Hafyani, Abdellah El Hmaidi, Abdelhadi El Ouali, Giovanni Randazzo and Ana C. Teodoro (2018). "Land Use/Land Cover (LULC) Using Landsat Data Series (MSS, TM, ETM+ and OLI) in Azrou Forest, in the Central Middle Atlas of Morocco." *Environments, 5*(12): 131.

Nongkynrih, A. Kyrham (2006). "Who is in? Who is Out? Equity and Customary Community Forest Management in Meghalaya, India." *Hanging in the Balance: Equity in Community-based Natural Resource Management in Asia*, pp. 47-62.

Ojima, D.S., K.A. Galvin and B.L. Turner (1994). "The Global Impact of Land-use Change." *Bio Science, 44*(5): 300-304.

Ormsby, Alison (2013). "Analysis of Local Attitudes Toward the Sacred Groves of Meghalaya and Karnataka, India." *Conservation and Society, 11*(2): 187-197.

Panigrahy, Rabindra K., Manish P. Kale, Upasana Dutta, Asima Mishra, Bishwarup Banerjee and Sarnam Singh (2010). "Forest Cover Change Detection of Western Ghats of Maharashtra Using Satellite Remote Sensing based Visual Interpretation Technique." *Current Science*, pp. 657-664.

Poffenberger, Mark (2014). "Khasi Responses to Forest Pressures: A Community REDD+ Project from North East India." *Forests under Pressure: Local Responses to Global Issues*, p. 229.

_______(2015). "Restoring and Conserving Khasi Forests: A Community-based REDD Strategy from North East India." *Forests 6*(12): 4477-4494.

Rai, Praveen Kumar, Rajeev Singh Chandel, Varun Narayan Mishra and Prafull Singh (2018). "Hydrological Inferences through Morphometric Analysis of Lower Kosi River Basin of India for Water Resource Management Based on Remote Sensing Data." *Applied Water Science, 8*: 1-16.

Ranjan, Avinash Kumar, Akash Anand, S. Vallisree and Rahul Kumar Singh (2016). "LU/LC Change Detection and Forest Degradation Analysis in Dalma Wildlife Sanctuary Using 3S Technology: A Case Study in Jamshedpur-India." *Aims Geosciences, 2*(4): 273-285.

Reddy, C. Sudhakar, C.S. Jha and V.K. Dadhwal (2013). "Assessment and Monitoring of Long-term Forest Cover Changes in Odisha, India Using Remote Sensing and GIS." *Environmental Monitoring and Assessment, 185*: 4399-4415.

_______(2016). "Assessment and Monitoring of Long-term Forest Cover Changes (1920-2013) in Western Ghats Biodiversity Hotspot." *Journal of Earth System Science, 125*: 103-114.

Sarma, Kiranmay (2005). "Impact of Coal Mining on Vegetation: A Case Study in Jaintia Hills District of Meghalaya, India." ITC.

Shalaby, Adel and Ryutaro Tateishi (2007). "Remote Sensing and GIS for Mapping and Monitoring Land-cover and Land-use Changes in the North-western Coastal Zone of Egypt." *Applied Geography, 27*(1): 28-41.

Silapaswan, Cherie Sumitra, D.L. Verbyla and A.D. McGuire (2001). "Land Cover Change on the Seward Peninsula: The Use of Remote Sensing to Evaluate the Potential Influences of Climate Warming on Historical Vegetation Dynamics." *Canadian Journal of Remote Sensing, 27*(5): 542-554.

Thiel, C., P. Drezet, C. Weise, S. Quegan and C. Schmullius. (2006). "Radar Remote Sensing for the Delineation of Forest Cover Maps and the Detection of Deforestation." *Forestry, 79*(5): 589-597.

Thomas, Nancy E., Chengquan Huang, Samuel N. Goward, Scott Powell, Khaldoun Rishmawi, Karen Schleeweis and Adrienne Hinds (2011).

"Validation of North American Forest Disturbance Dynamics Derived from Landsat Time Series Stacks." *Remote Sensing of Environment 115*(1): 19-32.

Tiwari, B.K., Saroj K. Barik and R.S. Tripathi (1998)."Biodiversity Value, Status and Strategies for Conservation of Sacred Groves of Meghalaya, India." *Ecosystem Health, 4*(1): 20-32.

Tiwari, B.K., H. Tynsong and M.B. Lynser (2010). "Forest Management Practices of the Tribal People of Meghalaya, North-East India." *Journal of Tropical Forest Science*, pp. 329-342.

Tiwari, B.K., H. Tynsong, M.M. Lynrah, E. Lapasam, S. Deb and D. Sharma (2013). "Institutional Arrangement and Typology of Community Forests of Meghalaya, Mizoram and Nagaland of North East India." *Journal of Forestry Research, 24*: 179-186.

Tiwari, L.K., S.K. Sinha and S. Saran (2016). *Forest Cover Change Detection in Andaman and Nicobar Islands Using Remote Sensing and GIS Techniques*, pp. 427-436.

Vashum, Kuimi Tampeimi (2016). "Impact of Slash-and-burn Cultivation on Floristic Diversity, Soil and Tribal Communities in Kamjong Sub-division, Manipur, India." Ph.D. dissertation, Department of Ecology and Environmental Sciences, Pondicherry University, Puducherry.

Yadav, Seema, Prodyut Bhattachrya, Deepakshi Babbar and Mayuri R. Wijesinghe (2022). "An Assessment of the Temporal Changes in Land Cover and Forest Fragmentation Using Geospatial Techniques: A Case Study from the Central Indian Highlands." In *Conservation, Management and Monitoring of Forest Resources in India*. Cham: Springer International Publishing.

Yin, He, Asia Khamzina, Dirk Pflugmacher and Christopher Martius. (2017). "Forest Cover Mapping in Post-Soviet Central Asia Using Multi-resolution Remote Sensing Imagery." *Scientific Reports, 7*(1): 1375.

Yismaw, Abyot, Birhanu Gedif, Solomon Addisu and Ferede Zewudu. (2014). "Forest Cover Change Detection Using Remote Sensing and GIS in Banja District, Amhara Region, Ethiopia." *International Journal of Environmental Monitoring and Analysis, 2*(6): 354.

Zhu, Zhiliang and David L. Evans (1994). "US Forest Types and Predicted Per Cent Forest Cover from AVHRR Data." *PE & RS-Photogrammetric Engineering & Remote Sensing, 60*(5): 525-531.

Section III
Environment and Agricultural Sustainability

11

Climate Smart Agriculture

An Approach to Transform Agricultural Systems for Food Security

Meghna Sarma

Introduction

Agriculture faces some stiff challenges ahead. It has to address the fact that almost one billion people go to bed hungry everyday, while more than two billion people will be added to the global population by 2050. In addition, food consumption patterns are changing as the average person in the world gets richer and consumes more food and more meat. There is increased competition for land, water, energy, and other inputs into food production. Climate change poses additional challenges to agriculture, particularly in developing countries. At the same time, many current farming practices damage the environment and are a major source (19-29%) of anthropogenic greenhouse gas (GHG) emissions. Farms emitted 6 billion ton of GHGs in 2011, or about 13 per cent of total global emissions. That makes the agricultural sector the world's second-largest emitter, after the energy sector (which includes emissions from power generation and transport). Most farm-related emissions come in the form of methane (CH_4) and nitrous oxide (N_2O). Cattle belching (CH_4) and the addition of natural or synthetic fertilizers and wastes to soils (N_2O) represent the largest sources, making up 65 per cent of agricultural emissions globally. Smaller sources include manure management, rice cultivation, field burning of crop residues, and fuel use on farms. Top ten countries with the largest agricultural emissions in 2011 were (in descending order): China, Brazil, United States, India, Indonesia, Russian Federation, Democratic Republic of Congo, Argentina, Myanmar, and Pakistan. Together, these countries contributed 51 per cent of global agricultural emissions. From 1990 to 2010, global agricultural emissions increased 8 per cent. They are projected to increase 15 per cent above 2010 levels by 2030, when they will amount to nearly 7 billion ton per year. These increases are mainly driven by population growth and changes in dietary preferences in developing economies. Agricultural emissions growth will be greatest in Asia and sub-Saharan Africa, which will account for two-thirds of the increase in overall food demand over first half of the 21st Century.

Climate-smart agriculture is a sort of overview concept originally put forth in 2010 by the UN's Food and Agriculture Organization. By climate smart, it means agriculture that sustainably increases productivity and resilience to environmental pressures, while at the same time reduces greenhouse gas emissions or removes them from the atmosphere, because we cannot ignore the fact that agriculture is itself a large emitter of greenhouse gases.

It is also known as Climate Resilient Agriculture (CRA). CRA means the incorporation of adaptation, mitigation and other practices in agriculture, which increases the capacity of the system to respond to various climate related disturbances by resisting damage and recovering quickly. Such perturbations and disturbances can include events such as drought, flooding, heat/cold wave, erratic rainfall pattern, long dry spells, insect or pest population explosions and other perceived threats caused by changing climate. CSA is an approach that has recently achieved much prominence, given the adaptation and mitigation challenges facing humanity. CSA is defined by three objectives: firstly, increasing agricultural productivity to support increased incomes, food security and development; secondly, increasing adaptive capacity at multiple levels (from farm to nation); and thirdly, decreasing greenhouse gas emissions and increasing carbon sinks. Since the relative priority of each objective varies across locations, with for example greater emphasis on productivity and adaptive capacity in low-input smallholder farming systems in least developed countries, an essential element of CSA is identifying potential synergies and trade-offs between objectives. CSA integrates climate change into the planning and implementation of sustainable agriculture and informs priority setting. In short, it is the ability of the system to bounce back. Climate resilient agriculture includes an in-built property in the system for the recognition of a threat that needs to be responded to, and the degree of effectiveness of the response. CRA will essentially involve judicious and improved management of natural resources, viz. land, water, soil and genetic resources through adoption of best bet practices.

Climate-smart agriculture sustainably helps in:

- Increases agricultural productivity,
- Builds resilience in food production systems, and
- Reduces greenhouse gas emission.

A variety of climate-smart practices already exist and are being used in some places, providing examples that could be more widely implemented in developing countries, as highlighted in an FAO report prepared in advance of the Cancun conference.

CSA practices and technologies adopted include:

- Improved crop varieties for higher yield,
- Varieties suitable to cope with drought,
- Excess water or high temperature,
- Laser land levelling,
- Practicing minimum tillage by using zero tiller or a happy seeder,
- Nutrient management by using green seeker,
- Managing irrigation by using tensiometer,
- Accessing weather information through SMS,
- Residue retention,
- Site specific nutrient management,
- Legume integration,
- Cropping system diversification,
- Use of solar pump,
- Use of crop sensor called a Green Seeker to assess crop health, and
- Use of a mobile phone app helps to calculate how much fertilizer to apply throughout the growing season.

History of Climate-smart Agriculture

If trends in human diet and waste in food systems remain unchecked, food production would have to increase by about 70 per cent to feed an estimated 9 billion people by 2050, with unprecedented consequences for the environment and society. The food price spikes of recent years have reinforced awareness of obvious links between political and economic stability and food security. Therefore, agricultural development is now the focus of renewed attention in both the research and policy communities. In describing tensions between maximizing global agricultural productivity, increasing resilience of agricultural systems in the face of climate change and mitigating greenhouse gas (GHG) emissions from agriculture, the term climate-smart agricultural development was first used in 2009. A year later, at the First Global Conference on Agriculture, Food Security and Climate Change at Hague, the concept of climate-smart agriculture (CSA) was presented and defined as agriculture that "sustainably increases productivity, enhances resilience, reduces/removes greenhouse gas emissions, and enhances achievement of national food security and development goals." This definition represented an attempt to set a global agenda for investments in agricultural research and innovation, joining the agriculture, development and climate change communities under a common brand. Drawing on this original framing, CSA has been applied to diverse aspects of agriculture, ranging from field-scale agricultural practices to food supply chains and food systems generally. Beyond agricultural practices and outcomes, a wide array of institutions, policies, finance, safety nets, capacity building and assessment have all been identified as enabling CSA. Following the Second Global Conference on Agriculture, Food Security and Climate Change in Hanoi in 2012, the recently published Climate-Smart Agriculture Sourcebook further advanced the concept with the intention of benefiting primarily smallholder farmers and vulnerable people in developing countries.

Need of CSA

The Indian agricultural production system faces the daunting task of having to feed 17.5 per cent of the global population with only 2.4 per cent of land and 4 per cent of the water resources at its disposal. With the continuously degrading natural resource base compounded further by global warming and associated climate changes resulting in increased frequency and intensity of extreme weather events, "business as usual" approach will not be able to ensure food and nutrition security to the vast population as well as environmental security (the need of the hour). The challenge is formidable because more has to be produced with reduced carbon and water footprints. To achieve this task of paving the way for climate smart agriculture we need to take several measures that will have enabling policies, institutions and infrastructure in place and the farming community be better informed and empowered with necessary resources. The world's population is expected to surpass 9 billion people by 2050, which will require an estimated 70 per cent increase in global agricultural production. At the same time, climate change is expected to have multiple impacts on agricultural productivity and rural incomes in areas that are already experiencing high levels of food insecurity. Climate change threatens to undo years of progress achieved in agricultural development, with the most severe consequences falling on developing countries and the poor. CSA can ensure agricultural development and food security are not compromised. Climate change simulation models indicate broad trends of warmer, wetter climate with more marked seasonality, and more unpredictable weather events that affect agriculture directly and indirectly. As a result, climate change is affecting crop yields, especially at low latitudes. Climate change is also affecting nutritional quality of food—C3 grains and legumes grown at elevated CO_2 concentrations have lower levels of zinc and iron. Climate change impacts on meat and fish production will also affect human and nutritional balance. Climate change is affecting marine fisheries—many species are moving away from the equator to cooler waters. Livestock too is affected—with decline in pasture productivity; this will lead to decline in meat and milk production. Given the large population in the Asia-Pacific region, with region-wide poverty, the number of people affected by climate change will be much higher than in other regions. To overcome the impact of climate change, an array of interventions for adaptation and mitigation has been tested. Specific adaptation measures include improved crop varieties and livestock and fish breeds, simple adjustments to land, crop and livestock management, more efficient use of water and energy for food production, better weather forecasts and timely provision of such data to farmers, and sustaining ecosystem services of forests and water sources. Likewise, mitigation of emissions or sequestering carbon can be achieved through soil management, livestock management, improved fertilizer management etc.

Overall, being climate-smart, as illustrated by the livestock sector, entails integrating various aspects of stock breeding and selection, animal health and disease management, grassland and feed management, and manure management. Practicing CSA is often considered a "no-regrets" strategy in that its benefits of sustainable improvements in agriculture will accrue even if climate change impacts turn out not to happen or be as severe as expected. Further, it is important to consider climate-smart management of agriculture within a broad socio-ecological context, recognizing that climate change is one of the many drivers of change influencing the agricultural sector. The CGIAR Research Programme (CRP) on Climate Change, Agriculture and Food Security (CCAFS) is piloting projects such as "climate-smart" villages to deal with climate change in a more holistic manner. The research is now being revised to address climate change as a cross-cutting issue to be addressed in meeting the challenges of reducing poverty, improving food and nutrition security, and improving natural resource systems and ecosystem services. Climate change is global, and response to its impacts must be universal.

Poor management practices in the rice-wheat cropping systems in the Indo-Gangetic Plains region for almost five decades have resulted in environmental degradation throughout the region. Today in North West India, combine harvesting of wheat is common practice (> 70% area), leaving large amounts (> 5 t ha-1) of crop residues in the fields. In the north-western states of Punjab and Haryana—once the heartland of the Green Revolution—farmers burn 30 million tons of crop residues annually, leading to "smoke, release of greenhouse gases, loss of plant nutrients and severe effects on human health", according to Rajbir Singh, Principal Scientist at ICAR who spoke at the event.

Concept of Climate Smart Village

The CSV approach was initiated first in Haryana and Bihar in 2011. Initially, 27 climate-smart villages from Nilokheri, Indri, Gharaunda and Nissing blocks in Karnal district of Haryana were brought under its ambit. Villages are the backbone of any development in India. Hence, any initiative taken, especially in the field of agriculture, must start from villages. In the modern scenario, the most discussed topics are of climate change and smart villages. It presents an integration of these two topics is relevant to a sustainable development. Climate smart village is a long-term concept that involves the interests of the local farmers, representatives and many other stakeholders. This feat can be achieved by introducing climate smart technologies, climate information services, local development, adaptation plans, supportive institutions and policies, all tailored to the community's needs. For the build-up of a climate smart village focus is on watershed management, digital technologies, meteorological data and climate based cropping models. A village is said to be smart when it is nutrient smart, water smart, crop smart, future smart, technology smart and energy smart. Nutrient smartness refers to the effective nutrient management through bio-fertilizer, bio-pesticides and integrated approach. Water smart village conserve the existing water resource by adopting practices like mulching and constructing of small ponds. Climate based cropping refers to crop smartness, where the crop is selected as per the prevailing conditions through crop rotation, mixed cropping and inter-cropping. Every farmer must be future smart by utilizing every scheme such as insurance for crop and livestock thereby mitigating loss, technical smartness is achieved by introducing simple, usable technologies such as SMS notification system which would provide information on weather, market prices etc. Above all, we should be energy smart as it is the core for sustainability. It can be achieved through bio gas plants, crop residue management etc. The road to achieve our goals in Agriculture is becoming difficult; going hand in hand with Mother Nature is the way further. Climate smart village is an idea towards that dream.

The location of a Climate-Smart Village is selected based on its climate risk profile and the willingness of farmers and local governments to participate in the project. Researchers conduct a baseline survey to capture the current socio-economic conditions and analyse resource availability and average production and income among other indicators. This enables an impact assessment after a period of time to gauge the benefits of the interventions. Stakeholders convene to prioritize which interventions they will take up that are best suited to their local conditions. This is done through a choice experiment that analyses their preference and willingness to pay for technologies. Disseminating information on Climate Smart Agriculture practices and outcomes is an important part of the capacity building process. Farmers are encouraged to record their testimonials and feedback at regular intervals and share it with researchers and the community.

Climate Smart Agriculture in India

Climate change and its variability are emerging as the major challenges influencing the performance of Indian agriculture.

Long-term changes in shifting weather patterns result in changing climate, which threaten agricultural productivity through high and low temperature regimes, increased rainfall variability, and rising sea levels that potentially deteriorate coastal freshwater reserves and increase risk of flooding. Climate change (and global warming) impacts all sectors of human life. Agriculture is particularly vulnerable to it. Each year one or the other part in the country is affected by droughts, floods, cyclones, hailstorm, frost and other climatic events. The fourth IPCC report clearly brought out the global and regional impacts of projected climate change on agriculture, water resources, natural eco-systems and food security. Among the several highly populated regions of the world, South Asia is categorized as one of the most vulnerable. Although climate change impacts are being witnessed all over the world, countries like India are more vulnerable in view of the huge population dependent on agriculture, excessive pressure on natural resources and poor coping mechanisms. India has already started development and inclusion of climate change adaptation policies in various sectors. The National Action Plan on Climate Change (NAPCC) of India identifies eight-core missions that promote various climate smart interventions in agriculture and allied sectors. For example, National Mission for Sustainable Agriculture (NMSA) and National Mission on Strategic Knowledge for Climate Change aim to support climate change adaptation in agriculture through the promotion of climate-smart practices and technologies across the country. State-level climate change adaptation plans in India focus on addressing the existing, as well as future challenges of climate change and take actions to reduce the associated risks and vulnerabilities.

The National Action Plan on Climate Change (NAPCC) of India

On June 30, 2008, Government of India released India's first National Action Plan on Climate Change (NAPCC) outlining existing and future policies and programmes addressing climate mitigation and adaptation. The plan identifies eight-core "national missions" running through 2017.

There are Eight National Missions, which form the core of the National Action Plan, representing multi-pronged, long-term and integrated strategies for achieving key goals in the context of climate change.

These missions are:

- National Solar Mission,
- National Mission for Enhanced Energy Efficiency,
- National Mission on Sustainable Habitat,
- National Water Mission,
- National Mission for Sustaining the Himalayan Ecosystem,
- National Mission for a "Green India",
- National Mission for Sustainable Agriculture, and
- National Mission on Strategic Knowledge for Climate Change.

National Initiative on Climate Resilient Agriculture (NICRA)

Indian Council of Agricultural Research, called National Innovations in Climate Resilient Agriculture, has now covered 151 villages across the country and plans to add another 100. NICRA is a network project of the Indian Council of Agricultural Research (ICAR) launched in February 2011 covers 151 villages across country. The project is formulated to take up long-term strategic research to address the impacts of projected climate change on Indian agriculture and demonstrate the existing best bet practices to enable farmers cope with current climate variability. Under the strategic research, phenol typing of germplasm collections of wheat, rice, maize and pulses is being done for climate-induced stresses like drought, heat stress and submergence, and selected lines are being used for developing stress tolerant high performing varieties. The project aims to enhance resilience of Indian agriculture to climate change and climate vulnerability through strategic research and technology demonstration. The research on adaptation and mitigation covers crops, livestock, fisheries and natural resource management. The project consists of four components, viz. Strategic Research, Technology Demonstration, Capacity Building and Sponsored/ Competitive Grants.

Dimensions of CSA

Weather-smart: Interventions that provide services related to income security and weather advisories to farmers:

- *Weather based Crop Agro-advisory:* Climate information based value added agro-advisories to the farmers.
- *Crop Insurance:* Crop-specific insurance to compensate income loss due to vagaries of weather.
- *Crop Diversification:* Growing a variety of crops instead of a single crop for income, diversifies a farmer's climate risk and contributes to increased household food security.
- *Climate Smart Housing for Livestock:* Protection of livestock from extreme climatic events (e.g. heat/cold stresses).
- *ICT based Agro-advisories:* Weather and climate information are being sent to subscribing farmers through SMS and voice regarding new seed varieties, climate smart farming practices and tips on conservation agriculture etc.
- *Climate Information:* Advance climate information help reduce climate risk or take advantage of better seasons.

Water smart: Interventions that reduce water requirements to produce the same or a higher level of yield:

- *Rainwater Harvesting:* Collection of rainwater not allowing to run-off and use for agriculture in rainfed/ dry areas and other purposes on-site.
- *Furrow Irrigated Bed Planting:* Effective control over irrigation and drainage as well as rainwater management during the monsoon.
- *Drip irrigation:* Application of water directly to the root zone of crops to minimize water loss.
- *SRI technology:* Low water, labour-intensive method.
- *Drainage:* Removal of excess water (flood) through water control structure.
- *Laser Land Levelling:* Levelling the field ensures uniform distribution of water and reduces water loss.
- *Direct Seeded Rice:* Delay in monsoon or water shortage, DSR gives the farmer flexibility to take up direct sowing of paddy with a suitable duration variety.
- *Cover Crops:* Reduces evaporation loss of soil water (also adds nutrients into the soil).

Carbon Smart: Interventions that reduce GHG emissions:

- *Agro-forestry:* Promote carbon sequestration including sustainable land use management.
- *Integrated Pest Management:* Reduces use of chemicals.
- *Fodder Management:* Promote carbon sequestration including sustainable land use management.
- *Livestock Management:* Concentrate feeding for livestock; Reduces nutrient losses and livestock requires low amount of feed.
- *Alternate Wetting and Drying:* Alternate wetting and drying management practices, which minimize flooding times, reduce both methane emissions and water use.

- *Bio-gas:* Reduced methane emission and fossil fuel free.

Nutrient and pest smart technologies: Technologies that save/supplement/avoid chemical fertilizer use for crops and control pests:

- *Site-Specific Nutrient Management:* Optimum supply of soil nutrients over time and space matching to the requirements of crops with right product, rate, time and place.
- *Green Manuring:* Cultivation of legumes in a cropping system. This practice improves nitrogen supply and soil quality.
- *Leaf Colour Chart/Green Seeker:* Quantify the required amount of nitrogen use based on greenness of crops. Mostly used for split dose application in rice but also applicable for maize and wheat crops to detect nitrogen deficiency.
- *Intercropping with Legumes:* Cultivation of legumes with other main crops in alternate rows or mixed. This practice improves nitrogen supply and soil quality.

Energy Smart: Technologies that help reduce energy consumption during land preparation without affecting yield levels:

- *Zero Tillage/Minimum Tillage:* Reduces amount of energy use in land preparation. In long-run, it also improves water infiltration and organic matter retention into the soil.
- *Use of Solar Pump:* Provide reliable, predictable and affordable energy inputs for irrigation.
- *Biochar/Biofuel:* Improve energy use efficiency, save fuel and improve indoor air condition, consequently saving time spent on fuel collection for women.

- *Mulching/Residue Management:* Maintains a protective cover over soil of crop residues and stubbles, improves infiltration, retains organic matter and water, improves the quality of seedbed.

Knowledge Smart: Interventions that use combination of science and local knowledge:

- *Use of Mobile Phone App:* Ensures pest and disease management of crops, fertilizer assessment etc.
- *Contingent Crop Planning:* Climatic risk management plan to cope with major weather-related contingencies like drought, flood, heat/cold stresses during the crop season.
- *Seed and Fodder Banks:* Farmers' platform for crop conservation/manage climate risks.
- *Gender Empowerment:* Engages women in different farm activities.
- *Rotational Grazing:* Sustainable fodder production from pastures.
- *Stress Tolerant High Yielding Breeds of Livestock:* Livestock breed that perform better under climatic stress/drought.

Technologies Adopted in CSV

- **Direct seeded rice:** Traditional rice cultivation involves sprouting rice in a nursery and then transplanting the seedlings into an intensively tilled field with standing water. With direct seeded rice, the rice seeds are sown directly in dry seedbed just like any other upland crop. This eliminates the laborious process of manually transplanting seedlings, significantly reduces the crop's water requirements and improves the soil's physical conditions.

- **Alternate wetting and drying in rice:** In alternate wetting and drying, rice fields are alternately flooded and drained. The use of a monitoring instrument like a tensiometer can help farmers decide when to irrigate their fields. Alternate wet and drying reduces methane emissions by an average of 48 per cent compared to continuous flooding. Combining this with precision fertilizer tools can further reduce greenhouse gas emissions.
- **ICT services to access weather and agro advisories:** M-Solution is CCAFS supported ICT-based climate and agro advisory project being piloted in Climate-Smart Villages by CIMMYT, together with Kisan Sanchar as the implementation partner and IFFCO Kisan Sanchar Limited (IKSL) as a content partner. Farmers get voice and text messages that inform them of weather forecasts, new seed varieties, climate smart farming practices and tips on conservation agriculture. The project aims to document farmers' perceptions on increasingly erratic weather events and to understand if the information they receive helps in overall behaviour change towards adapting to climate change and in the uptake of new practices and technology. The project ensured the inclusion of women farmers right from the onset. Many have said it is a vital source of information for them on climate change and agriculture.
- **Zero-tillage:** Zero-till or no-till farming is a way of growing crops without disturbing the soil through tillage. It increases the amount of water that infiltrates into the soil and increases organic matter retention and the cycling of nutrients in the soil. Zero-tillage improves soil properties, making it more resilient. A

CCAFS-CIMMYT study comparing zero-tillage and conventional tillage in Haryana showed that zero-tillage provided both economic and climate gains. Results show that farmers can save approximately USD 79 per hectare in terms of total production costs and increase net revenue of about USD 97.5 per hectare under zero-tillage land compared to conventional tillage. The study shows that shifting from conventional tillage to zero-tillage based wheat production reduces greenhouse gas emissions by 1.5 Mg CO_2-eq per hectare per season (Aryal *et. al.*, 2014). Zero-tillage also helps in buffering the terminal heat effects, which is one of the key climate change related constraints in wheat production in Haryana.

- **Laser land levelling:** A laser-leveller is a tractor-towed, laser-controlled device that achieves an exceptionally flat surface. Levelling the field ensures equitable reach and distribution of water and increases crop productivity. It also increases energy efficiency as less water means less need to run electric pumps. Most farmers in Haryana rent the equipment on an hourly basis. Farmers pay between Rupees 600-700 an hour to use the machinery. The practice of laser-land levelling has increased exponentially in the state. In Haryana, at the current adoption scale, the estimated amount of irrigation water saving is 933 million cubic meters per year. The estimated greenhouse gas mitigation is 1,63,600 MT of CO_2-eq per year (CIMMYT-CCAFS, 2014).
- **Residue management/mulching:** Crop residue mulching is a system of maintaining a protective cover of vegetative residues and stubble on the soil surface. It adds to soil organic matter, which improves the quality of the seedbed and increases the water infiltration and

retention capacity of the soil. Rice crop residue burning is one of the major issues in Haryana. Using innovative planting machinery like Turbo Happy Seeder, crops can be directly drilled without tillage while residue on surface acts as mulch.

- **Crop diversification:** The aim of crop diversification is to increase farmers' crop portfolio so they are not dependent on a single crop for an income. This also diversifies a farmer's climate risk and contributes to increased household food security. In addition, the innovative crop portfolio and system optimisation based diversification options help in sustainable intensification. Introducing legumes in the crop rotation cycle helps fix Nitrogen in the soil.
- **Agro-forestry:** Planting trees with crops or vegetables helps sequester carbon in the soil and prevent soil erosion. Trees provide shade to crops and are a source of timber, fruit, fodder and fuel-wood for farmers. They enhance biodiversity by providing habitats to varied species of birds, insects and animals and contribute to a healthy ecological landscape.

Success Stories

1. *Horizontal Expansion of Toria at NICRA Village*

Problem Identified: Lezai Miripathar Gaon and Janjimukh village, located near the river Burhidihing, under Barbarua Block of Dibrugarh district are chronically affected by flood every year. The water of Burhidihing, Sesa River and the backflow of mighty Brahmaputra inundate both villages. In recent times, the village experienced major flood in 2002 and 2016 with severe loss to crop and livestock. Recurrent flood not only causes damage to winter rice but also delays the sowing of Toria resulting in its low production. On the other hand, production of local cultivars of Toria being traditionally cultivated here is also low.

Technology injected: Demonstration on high yielding Toria TS-36 through its introduction in 2014-15 in 26 ha area using seed treatment with bio-fertilizers Azotobacter and PSB as a part of INM practice (75% RD of fertilizers and 25 per cent bio-fertilizers). The demonstration was conducted in four locations covering 46 farmers by sowing on 12-14th November, 2014. Detail is shown below:

Advantage of TS-67: This variety has wider sowing window. Normal-sown Toria variety (TS-36, TS-38) cannot be sown in December. However, this variety can be sown up to second week of December after harvesting of winter rice with comparatively higher yield and thereby helps in double cropping and increasing cropping intensity too. In 2016-17, in another 10 ha area TS-67 was demonstrated by KVK for upscaling of this variety in rice Toria sequence. By 2016-17 about 100 ha area in these two villages was covered under Toria cultivation. To show its replication, a programme was organized in these villages on 23 December, 2016 under "Technology Week" celebrated from 23 to 30 December, 2016 at different places of Dibrugarh district inviting different line departments and was telecasted through AIR and DDK, Dibrugarh. Yellow revolution in real sense has taken place in these two villages within three years.

Table 11.1: Performance of Toria var in Farmers field

Sowing time	Before intervention	After intervention
Toria from 15th Oct-15th Nov	8.00 q/ha	14.18 q/ha with TS-36 & TS-38
Toria up to 2nd week December	7.00 q/ha	10.25 q/ha with TS-67

Fig. 11.1: Performance of Toria var in Farmers field.

2. *Promotion of HYV of Sali paddy "Ranjit"*

The productivity of Sali paddy in West Garo Hills is very low as compared to other parts of the country due to non-availability of improved variety and non-application or hesitation to apply chemical fertilizers and method of cultivation. KVK, ICAR Tura has been selected by 1012 numbers of paddy growers in West Garo Hills as well as South Garo Hills district of Meghalaya for promotion of HYV of Sali paddy (Ranjit) during Kharif 2012 under TSP programme. Conducted trainings and field day also.

Fig. 11.2: Promotion of HYV of Sali paddy.

Table 11.2: Promotion of HYV of Sali paddy

Particulars	Yield(q/ha)	Cost of cultivation (Rs./ha)	Net Return(Rs./ ha)	C:B ratio
Ranjit	41.37	19,970	29,674	2.49
Local	25.70	15,540	10,160	1.65

Horizontal Expansion: The farmers who have grown Sali paddy variety Ranjit got bumper yield than other farmers. The farmers of neighbouring villages were encouraged automatically by "seeing is believing". About 250 ha area had been covered under cultivation of paddy variety, i.e. Ranjit in the district.

3. *SRI Paddy—Zero Tillage Greengram* (*var. Samrat*)

To address these problems of water scarcity and less production, resource conservation techniques of SRI methods of paddy cultivation using Ranjit variety was demonstrated in the NICRA villages. SRI method was introduced in five farmers field covering 1.5 ha area to 15 farmers. The average yield obtained by using SRI (46.53 q/ha) which was more than double as compared to traditional method of cultivation. These systems are gradually attracting farmers' attention and more and more farmers are approaching the KVK to help them to demonstrate on their farm to cope with natural calamities like water stress, terminal drought due to climatic variability. After the harvest of Sali paddy, few pockets of farmers field were demonstrated with short duration (55 days) greengram variety Samrat to increase the income during off season (*Rabi*) rather than traditional fellow land and obtained 3.83 q/ha yield. The REY of 59.30 q/ha was recorded with SRI paddy-Zero tillage greengram leads to gave maximum benefit of 2.44 compared to traditional paddy-fellow system.

Table 11.3: Performance of SRI practice followed by zero tilled green gram

RCT based Innovations	SRI paddy yield (q/ha)	Green-gram yield (q/ha)	System productivity (REY q/ha)	Net return (Rs./ ha)	Benefit cost ratio
SRI paddy-Zero tillage greengram	46.53	3.83	59.30	52450	2.44
Farmers' Practice (Rice-fallow)	22.78	–	22.78	10836	1.66

Fig. 11.3: Performance of SRI practice followed by zero tilled green gram.

4. *World's First Solar Irrigation Cooperative in Dhundi Village, Gujarat*

In May 2016, the world's first Solar Pump Irrigators Cooperative Enterprise (SPICE) commenced operations in the village of Dhundi. The International Water Management Institute (IWMI), which launched the project, demonstrated through

IWMI under the initiative of CCAFS. Six farmers formed the cooperative in February 2016.

Solarizing 1 Giga-watt capacity of electric irrigation pumps through the Dhundi approach will:

- Generate 1500 million units of green power annually.
- Put an additional income of more than USD 59 million into the hands of the farmers.
- Avoid groundwater draft of 1.7 billion cubic metres.
- Reduce CO_2 emissions by 1.23 million metric tons.
- Reduces land footprint of solar expansion: solar panels installed at an elevation allows to cultivate shade-loving crops below the panels.

Fig 11.4: Solar Panel at Dundi Village, Gujarat.

Case Studies

1. *Floating Gardens: Spreading the Ttradition*

The impact of climate change is seen vividly in Bangladesh. The cumulative effect of heavy rains, frequent storms and sea level rise is severe flooding. With continuous waterlogging conditions, crops are often lost and land for agriculture is becoming scarce. The low-lying areas of the southern coastal areas of Bangladesh remain submerged for long periods (generally from the month of June to December) every year, especially during the monsoon season. As a result, no crop cultivation is possible in this land naturally. The detrimental impacts of changing climate may intensify these situations in the future. In these circumstances, location-specific adaptation measures to climate change should be taken on a priority basis for improving the food security of the vulnerable people. The floating gardens/agriculture is based on local knowledge, locally as *dhap* or *baira* for their livelihood.

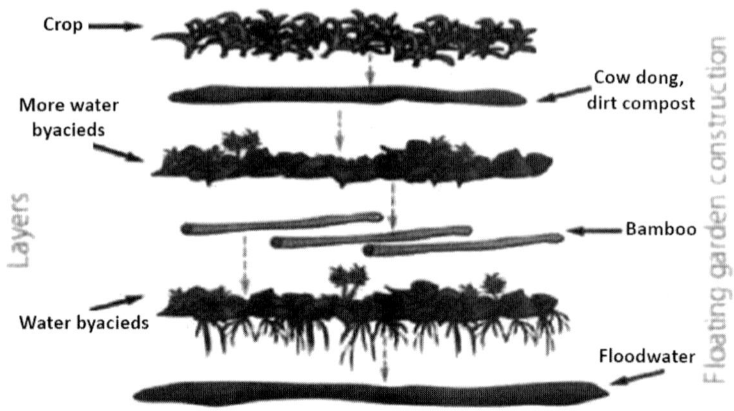

Fig 11.5: Construction of a Floating Garden.

These are floating plots made from local material on which diversified vegetables are grown or seedlings are raised for marketing. These plots resting on mats of plant material and bamboo can rise and fall with the river water levels, and are not washed away no matter how long the floods last. The floating agriculture is based on local knowledge, and is helping people to achieve sustainable livelihoods despite the adverse impacts of climate change.

Socio-economic Benefits

- Enhancement of nutritional security.
- Increase in quality food production.
- Increase in household income.

- Facilitation of employment in the wetlands.
- Promotion of a local technique.

Agricultural Benefits

Additional space for vegetables cultivation and seedling raising:

- Early production of seedlings of winter vegetables.
- Increase in vegetables supply in the area and the surroundings.
- Crops require shorter time to mature.
- Prime nutrient elements, namely N, P & K are available in water hyacinth comparable with cow-dung.

Fig. 11.6: View of Floating Gardens in Bangladesh.

2. *Preserving the Agro-forestry System on Mount Kilimanjaro (FAO, 2013)*

The "Kihamba" agro-forestry system covers 1,20,000 hectares of Mount Kilimanjaro's southern slopes. The 800 year-old system stands out among agro-forestry systems as one of the most sustainable forms of upland farming. Without undermining sustainability, it has been able to support one of the highest rural population densities in Africa, providing livelihoods for an estimated one million people.

As an ecologically compatible cash crop, coffee allowed the agro-forestry system to adapt successfully to the emerging cash economy. However, in the 1990s, coffee prices on the world market plummeted at the same time as pests and diseases were increasing. Additionally, many coffee shrubs had reached an age over 50 years when they produce less beans. These factors led to a sharp decline in productivity and profitability. It is estimated that 20 per cent of coffee cultivation in the area have been abandoned. If this continues, it will have massive environmental and socio-economic implications in the landscape around Mount Kilimanjaro especially on food security, carbon storage, water catchment and soil erosion. Under FAO's Globally Important Agricultural Heritage Systems Initiative (GIAHS), activities were piloted in 660 households to enhance farmers' cash income while preserving the ecological and social integrity of the Kihamba system. The project implemented an action plan, formulated together with the community, with the following key activities:

> Rethinking sources of cash income, three interventions were agreed on (a) conversion to certified organic coffee farming; (b) introduction of vanilla as a high value additional cash crop; and (c) introduction of trout aquaculture along the canals of the irrigation system. Also, rehabilitation of the

irrigation system to reduce water loss and expansion of the capacity of storage ponds to cope with longer dry seasons due to climate change, training in sustainable land management, the interventions in coffee management alone are expected to increase farm cash income by 25 per cent in three years.

3. *Climate Resilient Technological Interventions to Ensure Food Security in Flood Affected Areas—An Experience from Nicra Village, Dhubri, Assam*

With changing climate, extreme events are likely to increase in the country in general and in the north-eastern region in particular resulting into more drought and floods. Floods during the entire *Kharif* season hinders in achieving achievable yield and production threatening food security of majority of small and marginal farmers of the state (85%). In such situations, prevailing in one-third of rice land in Assam and elsewhere in the country may have specific technological interventions as has been demonstrated and proven in the present project villages of Dhubri district of Assam. To enhance the resilience to agriculture in flood affected and submergence prone area, high yielding rice varieties with certain specialities like 'Swarna Sub-1' and 'Luit' during *Kharif* and 'Joymati' Ahu, could be introduced which will certainly be adopted by farmers as per their agro-ecological situations.

Table 11.4: Impact and economics of submergence tolerant rice variety

Variety	No. of farmers	Area (ha)	Yield (q/ha)	Gross cost (Rs./ ha)	Gross return (Rs./ ha)	Net return (Rs./ ha)	B:C
Swarna sub-1	13	4	38.60	20,535	38,607	18,072	1.90
Traditional variety	13	4	26.54	18,342	26,538	8,196	1.44

Zero Till Wheat Cultivation Using Happy Seeder

In Rasidpur village of Ropar, wheat sowing with happy seeder technology provides an alternative to burning rice residues and allows direct drilling of wheat in standing crop residues. Initially, the happy seeder in adopted village was demonstrated on 5.8 ha and now it has increased up to 20 ha in 2015-16. A net saving of Rs. 4,735 per hectare was achieved when wheat was sown with happy seeder over conventional method.

Table 11.5: Economics of wheat sown with happy seeder and conventional method

Particulars	Method of sowing	
	Happy seeder	Conventional
Cost of field preparation (Rs./ha)	2500	6350
Weedicide and labour charges	400	1375
Rodenticide (Rs./ha)	90	–
Total (Rs./ha)	2990	7725
Net Saving (Rs./ha)	4735	–

Fig. 11.7: Use of Happy seeder for wheat sowing.

Conclusion

(i) Climate-smart agriculture addresses the complex interrelated challenges of food security, development and climate change, and identifies integrated options that create synergies and reduce trade-offs.

(ii) It assesses the interactions between sectors and the needs of different stakeholders.

(iii) It identifies barriers to adoption (esp. for farmers), and provides appropriate solutions in terms of policies, strategies, actions and incentives.

(iv) It seeks to create enabling environments through a better alignment of policies, investments and institutions.

(v) CSA strives to achieve multiple objectives with the understanding that priorities need to be set and collective decisions made on different benefits and trade-offs.

(vi) It prioritizes the strengthening of livelihoods (esp. those of smallholders) by improving access to services, knowledge, resources (including genetic resources), financial products and markets.

(vii) Developing countries must undergo a significant transformation in order to meet the related challenges of food security and climate change.

(viii) Adopting an ecosystem approach, working at landscape scale and ensuring intersectional coordination and cooperation is crucial for effective climate change responses.

(ix) Hence, identifying the key implementation strategies by linking climate smart policies with agricultural policies, it can be an effective tool to recognize and retain the skills, knowledge and resources of the rural economies, thereby addressing climate change across various agro-ecological zones.

REFERENCES

Amin, A., Mubeen, M., Hammad, H.M., Nasim, W. (2015). "Climate Smart Agriculture: An Approach for Sustainable Food Security." *Agric. Res. Commun*, *2*(3): 13-21.

Campbell, B.M., Thornton, P. Zougmore, R., Asten, P.V. and Ipper, L. (2014). "Sustainable Intensification: What is its Role in Climate Smart Agriculture." *Current Opinion in Environmental Sustainability, 8*: 39-43.

Deka, B.C., Deka, C.K., Das. P., Goswami, J. and. Bhattacharryya, H.C. (2017). "Climate Resilient Technological Interventions to Ensure Food Security in Flood Affected Area." *International J. of Plant Protection, 10*(2): 442-447.

FAO (2009a). *Food Security and Agricultural Mitigation in Developing Countries: Options for Capturing Synergies*. Rome, Italy.

FAO (2010). *Climate Smart Agriculture: Policies, Practices and Financing for Food Security, Adaptation and Mitigation*. Rome, Italy.

FAO (2014). *Transitioning towards Climate-smart Agriculture in Kenya*. Food and Agriculture Organization of the United Nations, Rome. http://www.fao.org/3/a-4259

Irfanullah, H., Kalam, A., Kamruzzaman, M. & Wahed, A. (2011). "Floating Gardening in Bangladesh: A Means to Rebuild Lives after Devastating Flood." *Indian Journal of Traditional Knowledge, 10*(1), pp. 31-38.

Khatri-Chhetri, Arun, Aggarwal, P.K., Joshi, P.K. and Vyas, S. (2017). "Farmers' Prioritization of Climate-Smart Agriculture (CSA) Technologies." *Agriculture Systems*, pp. 184-191.

Kumar, K., Hanumanthappa, M., Narayan Mavarkar, S. and Marimuthu, S. (2018). "Review on Smart Practices and Technologies for Climate Resilient Agriculture." *International Journal of Current Microbiology and Applied Sciences, 7*(6): 3021-3031.

Maheswari, M., Srinivasarao, M., Rao., V.U.M., Srinivasarao, CH., Reddy, K.S., Ramana, D.B.V., Ramana Rao, C.A., Vijay Kumar, P., Prasad, Y.G., Sikka, A.K., Venkateswarlu, B. (2014). *National Initiative on Climate Resilient Agriculture (NICRA), Research Highlights*. Central Research Institute for Dryland Agriculture (ICAR), Hyderabad.

Prasad, Y.G., Maheswari, M., Dixit, S., Srinivasarao, Ch., Sikka, A.K., Venkateswarlu, B., Sudhakar, N., Prabhu Kumar, S., Singh, A.K., Gogoi, A.K., Singh, A.K., Singh, Y.V. and Mishra, A. (2014). *Smart Practices and Technologies for Climate Resilient Agriculture*. Central Research Institute for Dryland Agriculture (ICAR), Hyderabad.

Shah, R. and Srivastava, R. (2017). "Effect of Global Warming on Indian Agriculture." *J. of Climatology & Weather Forecasting, 5*:1.

12

A Sustainable Approach to Circularize Effluent Management in Dairy Industry

A Comprehensive Review

Sushma Gautam, Jyoti Sharma, Sonali Bhambhani, Nandini Bansal, Deepal Sharma, Daksh Massey, Nikhil Dogra and *Neha Sharma*

Introduction

Heightened consumer demands stem from an expanding population and urbanization. The mounting pressures on limited resources of the earth has accelerated the destruction of environment and made the climate footprints much wider (Kilic *et al.*, 2021). Recycling of resources is, therefore, of paramount importance for the future healthy life of the planet. Water supply and wastewater treatment, two most important indicators of our sustenance, are continuously under peril, looking into dwindling water supplies and release of untreated wastewaters are causing eco-toxicological implications. One of the reasons for decreased fresh water supply is because the capacity of the natural ecosystems to dilute the anthropogenic contaminants has been diminishing attributable to paucity of treatment facilities. Deterioration of water quality and availability has become the major threat to environmental sustainability (Voulvoulis, 2018).

Water scarcity is a global concern, where the demand approaches or exceeds the fresh water supply (Makropoulos *et al.*, 2018). The sixth goal of the 2030 Sustainable Development Goals (SDGs) is *Ensure availability and sustainable management of water and sanitation for all* (UN). Some of the specific targets can be connected to implementation of CE. *Treatment and reuse of wastewater and ambient water quality* is one of the targets of SDG-6, which is monitored by UN Environment through a cohesive approach facilitated by member countries. The issues that are identified for the agenda of SDGs such as raising resource inequalities, natural resources depletion, environmental degradation and climate change, are in conjunction with water resources (WWAP, 2017). A vast majority of wastewater remains partially or completely untreated and finds its way into adjoining surface waters (Sharma *et al.*, 2020). Water bodies contaminated with diversified recalcitrant moieties pose a significant threat to both biotic and abiotic resources. By processes like *bio-accumulation* and *bio-magnification*, these toxicants contribute to potential biohazards through their circulation across different trophic levels. With anticipated financial depreciation and addressing the *pressure*, linear economic models have somehow necessitated eco-friendly, sustainable viable, cost-effective solutions for catapulting economic trends in allied water sectors. This perhaps has necessitated emergence of CE (Reichel *et al.*, 2016).

Waste and CE: An Anticipated Correlation

In terms of CE, waste is identified as a resource, which is generally misplaced, therefore, accurately referred to as "by-product" instead of "waste" (Kirchherr *et al.*, 2017). With unprecedented rise in unsustainable growth of mankind, the natural ecosystem, which has measures in place to ensure a no-waste cycle of material, has been exchanged for an artificial man-made ecosystem by the intensification of anthropogenic activities. All the biotic and abiotic factors are responsible for generation of waste in some or the other manner. Thus, waste is a consequence of everyday life—of all creatures (Worrell *et al.*, 2016). All verticals of nation's economy—farms, mines, and factories as well as businesses, institutions, and households —contributes to the mounting mass of unwanted materials requiring disposal. It is inclusive of household waste, separately collected—bulky waste, bio-waste, market waste, road-sweeping waste, paper and glass (Rathoure and Dhatwalia, 2016). Thus, there is a necessity of streamlining the heaps of waste using sustainable tactics by CE approach. Schematic representation of LE *versus* CE is depicted in Figure 12.1.

The general economic production patterns that are predominant in the present world are based on simple linear processes—extract raw materials to make finished products, then consume them and generate waste. The thrust is given to the economic objectives, with least regard to the environmental repercussions. In CE, instead of entirely depending on the virgin resources, byproducts, wastes or discarded products are used as a primary resource for production of new product. This would consequently reduce the generation of wastes in each of the manufacturing process (STOP, 2017). The European Commission (EC) in 2015 has come up with a definition of CE which is as follows: "*CE is a system where the value of products, materials and resources is maintained in the economy for as long as possible, and the generation of waste is minimized*" (CEC, 2015). To let a feasible model prosper, different industries would combine their production loops synergistically so that the byproducts and wastes of one industry become resources

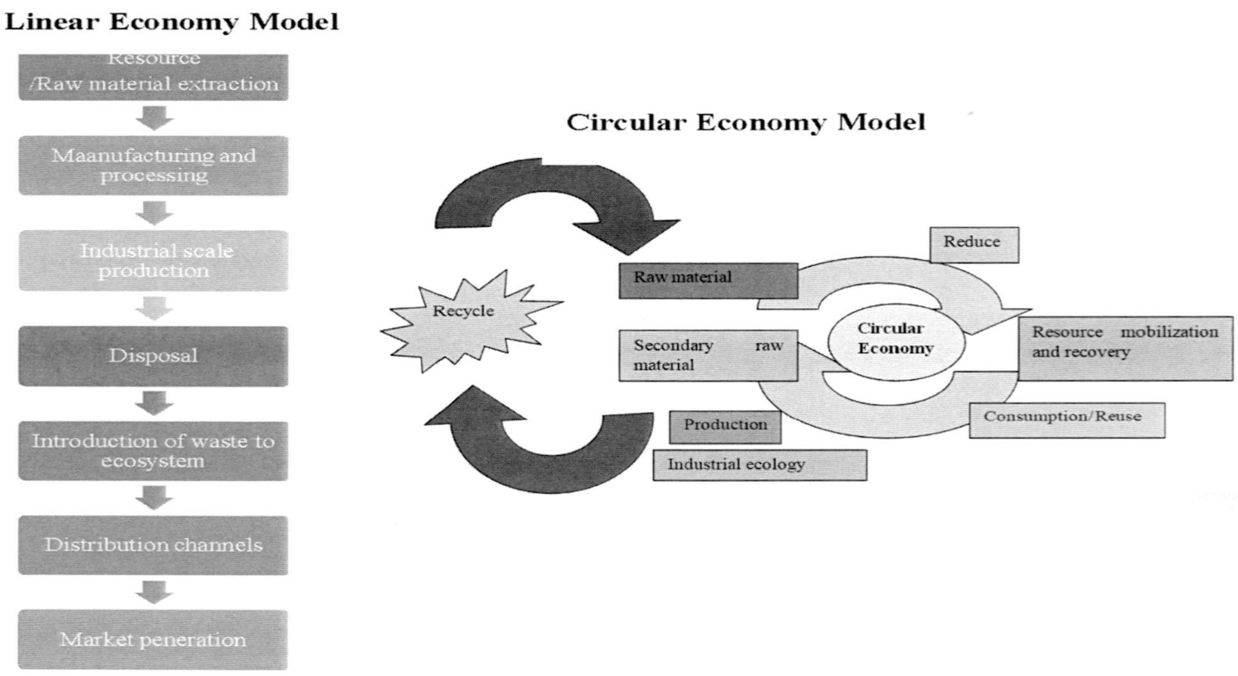

Fig. 12.1: Contrasting characteristics between LE and CE.

for another. The energy and other resources are also shared among these industries in such a symbiotic setting so that there are optimum resource utilization and waste assimilation (Smol *et al.,* 2020).

There are compelling benefits in moving from a LE to a CE in terms of environmental, economic and societal benefits. Due to lower carbon emissions, lower costs, and decreased supply chain risk, the CE is predicted to save over $1 trillion annually globally. Additionally, it can safeguard about half trillion dollars of India's GDP value (Fiksel *et al.,* 2021). According to the present growth scenario, India might produce an annual value 14 trillion INR and 40 trillion INR in the field of CE by 2030 and 2050, respectively. Many researchers assessed the circularity degree (Fig. 12.2) and impact of circularity (Fig. 12.3) using numerous tools (Bherwani *et al.,* 2022).

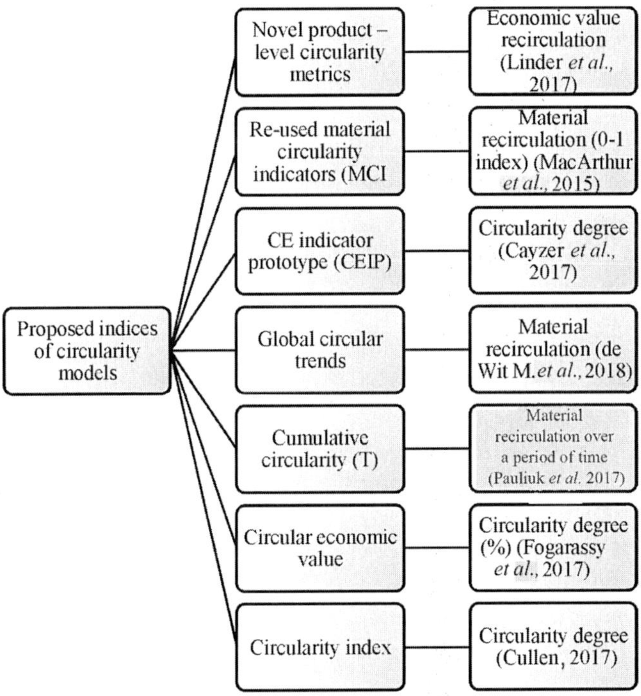

Fig. 12.2: Circularity indices.

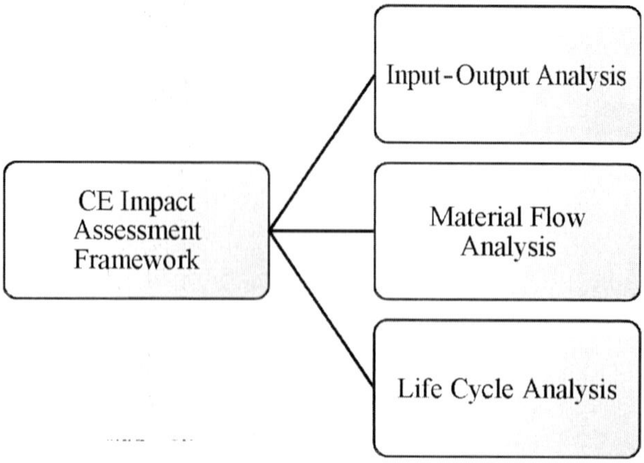

Fig. 12.3: Tools Used to Analyze the Resultant Impact of a CE.

The evolution of CE is, therefore, important for environmental sustainability, in which, economic development is equilibrated by protection of natural resources and environmental sustainability. In January 2016, the UN proclaimed SDG-2030, aimed towards reforming sustained, inclusive and sustainable economic growth, in global context. Some of the goals are directly related to the use of water and treatment of wastewater; a paradigm shift into CE would certainly accelerate the attainment of many of the SDG 2030 (IWA, 2016) (Fig. 12.4).

Transitioning of the water consuming industries to the CE contributes to an economic sustenance *via* technological inputs leading to an enhanced efficacious unit processes and operations. With upgradation in current treatment regime, impetus has been given to resource recovery. "End-of-pipe' treatment strategy has been practiced globally to treat sector specific influents with maximum possible nutrient recovery in addition to conventional approach (Sharma *et al.*, 2021).

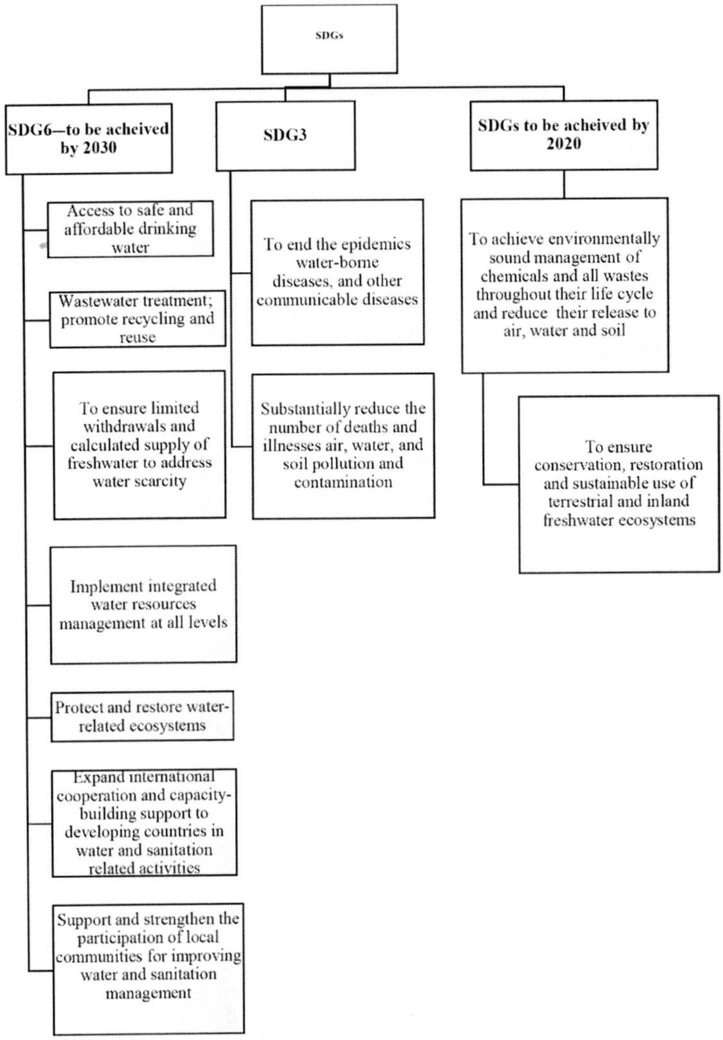

Fig. 12.4: SDGs and their deliverables.

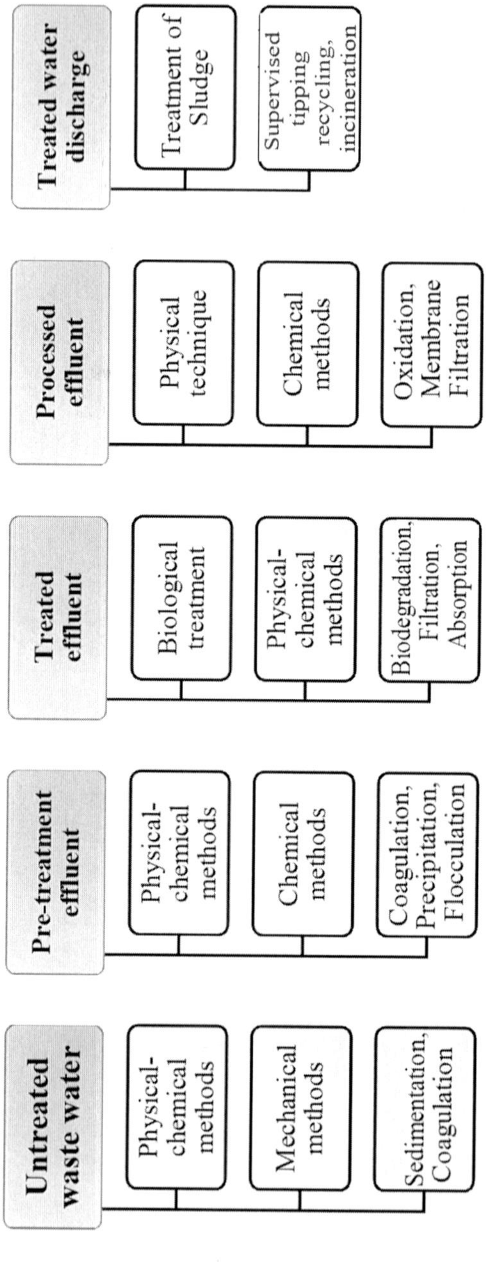

Fig. 12.5: General outlay of mitigating contaminants from waste water and unit operations.

Sector Specific Challenges: Industrial and Domestic

Ever-increasing water scarcity issues are of utmost concern requiring immediate address. The water quality is deteriorated mainly by discharge of untreated wastewater by municipal and industrial sectors. The industries contributing to addition of contaminants to water bodies includes food processing, agro-industries, tanneries, textile and paper industries etc. Current wastewater treatment methods involve a combination of physical, chemical and biological processes and operations to remove insoluble particles and soluble contaminants from

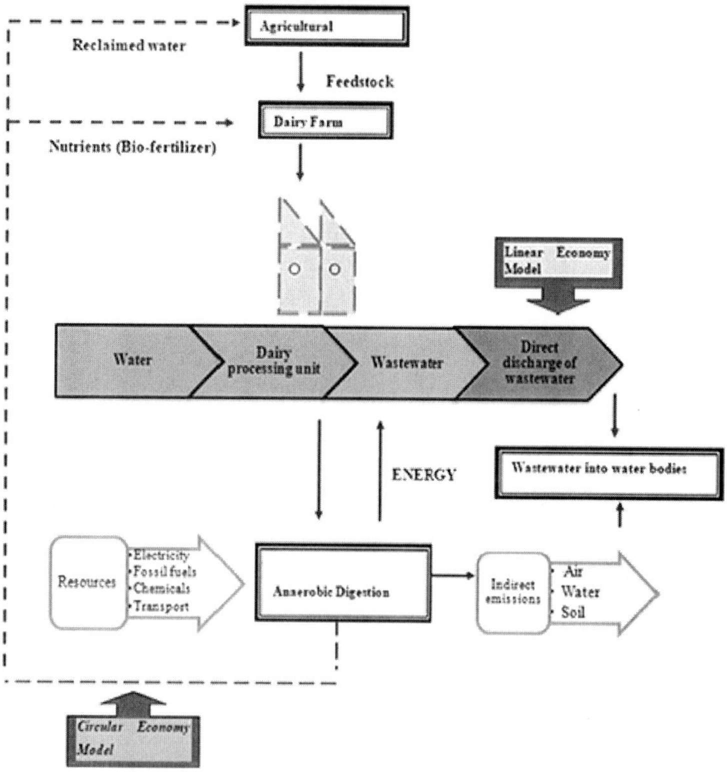

Fig. 12.6: Lucid overview of advancement of wastewater treatment method.

effluents. Figure 12.5 depicts the basic process of treating polluted water in sewage treatment plants and effluent treatment plants (Sharma *et al.,* 2013).

During the course of societal development, the processes of treating effluent rich water had been changed dynamically. The aim behind transitioning of technology is the development of a more sustainable, cost-friendly and eco-effective way of removing pollutants from the ecosystem. Figure 12.6 gives a lucid overview of advancement of wastewater treatment methods along few decades (Morin-Crini *et. al.,* 2022).

Dairy Industry and Circularity with an Essence of Life Cycle Impact Assessment (LCIA)

Referring to agro-industrial and food processing agrarian practices, dairy sector has captured attention, lately. Effluent management systems undergo conventional treatment practices, encompassing *three tier strategic* approach without *"circularizing"* effluent release. A conventional and linear agribusiness model constitutes the following unit operations (Kirilova and Vaklieva-Bancheva, 2017):

- Production,
- Processing,
- Manufacturing,
- Quality assessment,
- Validation,
- Dispatch,
- In-house treatment,
- Marketing,
- Cash flow, and
- End user feedback.

Globally, dairy industries are categorized as prominent users of water resources, thereby releasing subsequent volumes of wastewater. The dairy effluent is enriched in nutrients such as solids ammonia, nitrite, nitrate, milk protein, fat, amino acid, urea, uric acid, trivial nitrogen ammonia salt and phosphorus (Sharma and Dwivedi, 2017). The high nutrient content in diary wastewater is the culprit behind causing *bio-accumulation* into biota and *eutrophication* of water bodies (Azman *et al.*, 2016). Various methods have been explored for treatment of dairy effluent including electro-coagulation process, phyto-remediation process, water hyacinth process, aerobic and anaerobic processes and reverse osmosis phenomenon and by some electrochemical processes (Ahmaruzzaman and Gupta, 2011). However, the proper treatment and management strategy for mitigating contaminants is still a challenge.

A *Life Cycle Impact Assessment* (LCIA) is a modern approach which delineates all the stages of production from beginning to the end into sub-processes, which is 'cradle to grave' study (Zijp *et al.*, 2017). The transportation between the industrialization and commercialization of dairy products; the waste streams pertaining to each stage are also part of the LCIA framework. This approach is useful to assess the environmental impacts of dairy industry and improve its sustainability. In this account, development of sustainable dairy valorizing chains should have implemented, considering the reduction of the ecological impacts and cause-effect relationships within all stages of the supply chain. LCIA can be employed as tool to adjudicate the decision-making in the dairy sector, in order to introduce environment-friendly amendments. Several studies have been carried out to assess the potential of dairy wastewater as a raw material to execute CE principles.

- Conversion of waste into Bioenergy (Kimming *et al.*, 2015).
- Nutrient recycling (Dolman *et al.*, 2014).
- Recovery of polyhydroxy-alkanoate (PHA) from permeate fraction containing lactose (Rosa, 2018) wastewater treatment and valorisation (Gottschall *et al.*, 2007).

The elaborated environmental impacts of dairy value chains including, Global Warming Potential (GWP), eutrophication potential, acidification potential, photochemical ozone formation, ozone depletion, energy use, resource use, land use, eco-toxicity, water depletion etc. The concept of CE in dairy industry focuses on reduction and re-use of resources and wastes with a focus on sustainability. Even though the dairy sector traditionally involves CE practices, there are still many miles to attain a circular dairy sector that is regenerative and with closed nutrient, water, carbon and waste cycles.

Robust nutrient re-cycling technology has been evolved during the recent past, nutrient conversion of dairy waste has become more economically viable in current times. Hence, nutrient recycling is a dominant process in dairy industry as a result of its environmental and economic benefits. Current *Anaerobic Digestion* (AD) technologies are capable of minimizing waste and retaining valuable materials and energy within the dairy value chains as dairy wastewaters have a significant organic and microbial load. However, optimization of the operating conditions poses a significant challenge in wide implementation of AD technologies in the dairy industry. Figure 12.7 reflects unit processes indigenous to dairy sector, their linearity and a scope of improvisation into a circular model (Stanchev *et al.*, 2020).

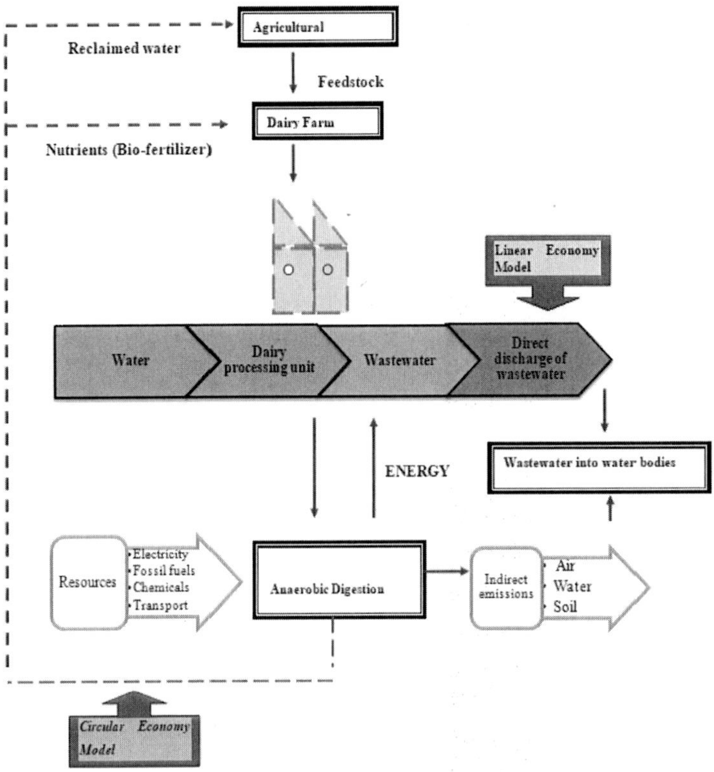

Fig. 12.7: Circular Economy Loop in Dairy Supply Chain.

Biotechnological Interventions for Resource Recovery

Biological interventions in resource recovery in recent times have carved a niche in extracting valuable resources during wastewater treatment. Indigenous microorganisms have been explored to their maximum potential for their bio-degradative pathways culminating into value-added product generation. This necessitates further explorative studies aimed at microbial ecology of wastewater (Saxena *et al.*, 2015). This has led to the emergence of a relatively new concept of metagenomics and transcriptomics.

Microbial technology has been used to generate useful biogas from animal and agricultural waste. Even though it is common now-a-days that biogas is used directly for lighting and as fuel, electricity generation from biogas is not yet common in most part of the world. Wastewater treatment plants (WWTP) can also generate electricity from biogas using the same anerobic digestion process.

Microbial activities have been employed for water purification for quite a long time. It has been demonstrated that indigenous microbes had the ability to lower the pollution indicators from toxic textile industry (Chanwala *et al.*, 2019). The study based on textile industry effluents from Sanganer town, Rajasthan to demonstrate the bio-absorption of synthetic dyes using attenuated *Enterobacter sp.* In their experiment with the textile effluents inoculated with acclimatized bacterial cells, pollution indicators such as colour, pH, BOD, COD, nitrate and phosphate shows considerable reduction in comparison with the same effluents without the inoculums (Sharma *et al.*, 2020).

Nielsen (2017) discusses about microbial technologies to remove and reuse C, P, N and micro-pollutants from wastewater and production of energy and byproducts thus making it an integral part of circular economy. The major challenge here is to manage the bacterial communities. MiDAS, a worldwide survey of microbes in wastewater treatment systems and anaerobic digesters has a summarized list of all these engineered microbial ecosystems.

Bio Electro Mechanical systems (BEM) use microbial metabolic processes to convert oxidizable organic matter into energy (bioelectricity) and other useful products. This technology has many advantages over the conventional AD, which include the ability to treat organic substrate even with low COD concentration, higher concentration of volatile fatty acids and the process produces less quantity of sludge (Robles *et al.*, 2020).

Microbial Fuel Cells (MFC) can achieve more efficient energy extraction from organics during the water treatment process, which can greatly reduce the energy required for the process (Gu *et al.*, 2017). However, there are many challenges yet to be addressed by future research: high cost, low power outputs are the main concerns. Further researches are also required to understand the metabolic mechanisms, improve the design and architecture of the reactors, address voltage reversals and ionic short circuits, develop better power collection and utilization systems, and streamlining with other wastewater technology that may accelerate their application in wastewater treatment.

Conclusion

CE being the most promising route towards sustainable growth focuses on material circularity implied that it must replenish the environment, economy, and society. This review assesses the existing circularity measures and indices. However, certain frameworks for evaluation, including LCIA (and derived indicators), have demonstrated significant promise for achieving all of the CE's objectives at the product and service levels. According to the literature study, LCA is the most popular framework for evaluating circular methods. With advent of biotechnology-enabled processes, an insight pertaining to resource segregation, recovery, utilization and mobilization has gained momentum in recent times. That being said, industrial ecology practices are being practiced to attain sustainable unit operations in industries leveraging into circularity. Circular economy thus promotes *Lifestyle For Environment* (LiFE), therefore, it is an effort in inching closer to *Cleaner, Greener and Bluer* future. The future sustenance using LiFE can be attained by responsible choices at both personal level and national development level. G-20 mandate 2023 has focused on the human-environment interaction with planet earth, hence the

Indian Presidency mainly emphasized on *"One Earth, One Family, One Future"*. Thus, the circularity of resources using biotechnological interventions is a preferred measure for future sustenance.

Acknowledgements

The authors would like to thank Department of Bio-Sciences and Technology, Maharishi Markandeshwar Engineering College, Maharishi Markandeshwar (Deemed to be) University, Mullana, Ambala (Haryana) for providing all necessary resources and access to library information systems for comprehensive literature search.

REFERENCES

Ahmaruzzaman, M. and V.K. Gupta (2011). "Rice Husk and Its Ash as Low-cost Adsorbents in Water and Wastewater Treatment". *Industrial & Engineering Chemistry Research, 50*(124): 13595. https://doi.org/10.1021/ie201802z

Azman, Nadia Farhana, Peyman Abdeshahian, Najeeb Al-Shorgani, and Mohd Sahaid Kalil (2016). "Production of Hydrogen Energy from Dilute Acid-Hydrolyzed Palm Oil Mill Effluent in Dark Fermentation Using an Empirical Model." *International Journal of Hydrogen Energy, 41*(37): 16373-16384. https://doi.org/10.1016/j.ijhydene.2016.05.085

Cayzer, S., P. Griffiths, and V. Beghetto (2017). "Design of Indicators for Measuring Product Performance in the CE". *International Journal of Sustainable Engineering, 10:* 289-298.

Chanwala, Jeky, Garima Kaushik, Mohd Ashraf Dar, Shivangi Upadhyay and Akhil Agrawal (2019). "Process Otimization and Enhanced Decolorization of Textile Effluent by *Planococcus sp.* Isolated from Textile Sludge." *Environmental Technology and Innovation 1*(13): 122-129. https://doi.org/10.1016/j.eti.2018.11.008

Commission of European Communities (2015). *Closing the Loop—An EU Action Plan for the CE. COM(2015) 614.*

Cullen, J.M. (2017). *CE: Theoretical Benchmark or Perpetual Motion Machine?.* Cambridge. https://api.repository.cam.ac.uk/server/api/core/bitstreams/37832832-7821-4eed-b36e-51e6643dc1af/content

Dolman, M.A., M.P.W. Sonneveld, H. Mollenhorst and I.J.M. de Boer. 2014. "Benchmarking the Economic, Environmental and Societal Performance of Dutch Dairy Farms Aiming at Internal Recycling of Nutrients." *Journal of Cleaner Production, 73:* 245-252. https://doi.org/10.1016/j.jclepro.2014.02.043

European Parliamentary Research Service (2017). *Towards a Circular Economy—Waste Management in the EU.*

Fiksel, J., P. Sanjay, and K. Raman (2021). "Steps Toward a Resilient CE in India." *Clean Technologies and Environmental Policy, 23:* 203-218.

Fogarassy, C., B. Horvath, A. Kovacs, L. Szoke, K. Takacs-Gyorgy (2017). "A Circular Evaluation Tool for Sustainable Event Management—An Olympic Case Study."*Acta Polytechnica Hungarica,14:* 161-177.

Gottschall, N., C. Boutin, A. Crolla, C. Kinsley, P. Champagne (2007). "The Role of Plants in the Removal of Nutrients at a Constructed Wetland Treating Agricultural (Dairy) Wastewater, Ontario, Canada." *Ecological Engineering, 29:* 154-163. https://doi.org/10.1016/j.ecoleng.2006.06.004

Gu, Yifan, Yue Li, Xuyao Li, Pengzhou Luo, Hongtao Wang, Zoe.P. Robinson, Xin Wang, Jiang Wu (2017). "The Feasibility and Challenges of Energy Self-Sufficient Wastewater Treatment Plants". *Applied Energy, 204:* 1463-1475.

Kilic, Z. (2021). "Water Pollution: Causes, Negative Effects and Prevention Methods".*Ýstanbul Sabahattin Zaim Üniversitesi Fen Bilimleri Enstitüsü Dergisi, 3*(2): 129-32.

Kimming, M., C. Sundberg, Å Nordberg, A. Baky, S. Bernesson, P.A. Hansson (2015). "Replacing Fossil Energy for Organic Milk Production—Potential Biomass Sources and Greenhouse Gas Emission Reductions". *Journal of Cleaner Production, 106:* 400-407, https://doi.org/10.1016/j.jclepro.2014.03.044

Kirchherr, J., D. Reike, and M. Hekkert (2017). "Conceptualizing the CE: An Analysis of 114 Definitions." *Resources, Conservation and Recycling, 127:* 221-232.

Kirilova, E.G. and N.G. Vaklieva-Bancheva (2017). "Environmentally Friendly Management of Dairy Supply Chain for Designing a Green Products Portfolio." *Journal of Cleaner Production, 167:* 493-504. https://doi.org/10.1016/j.jclepro.2017.08.188

Linder, M., S. Sarasini, and P. van Loon (2017). "A Metric for Quantifying Product Level Circularity". *Journal of Industrial Ecology, 21*(3): 545-558.

MacArthur, E., K. Zumwinkel, and M.R. Stuchtey (2015). *Growth Within: A Circular Economy Vision for a Competitive Europe.* Ellen MacArthur Foundation.

Makropoulos, C., E. Rozos, I. Tsoukalas, A. Plevri, G. Karakatsanis, L. Karagiannidis, E. Makri, C. Lioumis, C. Noutsopoulos, D. Mamais, C. Rippis, E. Lytras (2018). "Sewer-Mining: A Water Reuse Option Supporting CE, Public Service Provision and Entrepreneurship." *Journal of Environmental Management, 216:* 285-298.

Morin-Crini, Nadia, Eric Lichtfouse, Marc Fourmentin, Ana Rita Lado Ribeiro, Constantinos Noutsopoulos, Francesca Mapelli, Éva Fenyvesi, Melissa Gurgel Adeodato Vieira, Lorenzo A. Picos-Corrales, Juan Carlos Moreno-Piraján, Liliana Giraldo, Tamás Sohajda, Mohammad Mahmudul Huq, Jafar Soltan, Giangiacomo Torri, Monica Magureanu, Corina Bradu & Grégorio Crini (2022). "Removal of Emerging Contaminants from Wastewater Using Advanced Treatments: A Review." *Environmental Chemistry Letters 20*(2): 1333-1375. https://doi.org/10.1007/s10311-022-01402-4

Nair, Moorthy, Amol Niwalkar, Dhanya Balachandran and Rakesh Kumar (2022). "Application of CE Framework for Reducing the Impacts of Climate Change: A Case Study from India on the Evaluation of Carbon and Materials Footprint Nexus." *Energy Nexus, 5:* 100047.

Nielsen, P.H. (2017). "Microbial Biotechnology and Circular Economy in Wastewater Treatment." *Microbial Biotechnology, 10*(5): 1102-1105. https://doi.org/10.1111/1751-7915.12821

Pauliuk, S., Y. Kondo, S. Nakamura and K. Nakajima (2017). "Regional Distribution and Losses of End-of-Life Steel throughout Multiple Product Life Cycles—Insights from the Global Multiregional MaTrace Model". *Resource Conservation and Recycling, 116:* 84-93. https://doi.org/10.1016/j.resconrec.2016.09.029

Rathoure, A.K. and V.K. Dhatwalia (eds.) (2016).*Toxicity and Waste Management Using Bioremediation.* Hershey. PA: IGI Global.

Robles, Ángel, Daniel Aguado, Ramón Barat, Luis Borrás, Alberto Bouzas, Juan Bautista Giménez, Nuria Martí, Josep Ribes, María Victoria Ruano, Joaquín Serralta, José Ferrer, Aurora Seco (2020). "New Frontiers from Removal to Recycling of Nitrogen and Phosphorus from Wastewater in the Circular Economy." *Bioresource Technology. 300:* 122675. https://doi.org/10.1016/j.biortech.2019.122673

Rosa, F. (2018). "Waste Generated by Food Industry and Reuse in a Circular Economy Approach: The Whey Processing." *Concepts of Dairy & Veterinary Sciences, 2:* 171-173. https://doi.org/10.32474/CDVS.2018.02.000132

Saxena, S., N. Sharma, and A.D. Gupta (2015). "Isolation and Characterization of Bacterial Strains in Terms of COD Reduction from Dairy Waste Water". *Journal of Environmental Science, Computer Science and Engineering & Technology, 4:* 319-328.

Sharma, N. and A. Dwivedi (2017). "Bioremediation of Dairy Waste Water for Nitrate Reduction." *World Journal of Pharmaceutical and Life Sciences, 3*(1): 375-84.

Sharma, N. and S.K. Sharma (2021). "Wastewater Treatment Plants as Emerging Source of Antibiotic Resistance". *Green Chemistry and Water Remediation: Research and Applications,* pp. 239-269.

Sharma, N., H. Bhagwani, N. Yadav, and D. Chahar (2020). "Biodegradation of Textile Wastewater by Naturally Attenuated Enterobacter sp." *Nature Environment and Pollution Technology, 19*(2): 845-850.

Sharma, N., S. Chatterjee, and P. Bhatnagar (2013). "An Evaluation of Physicochemical Properties to Assess Quality of Treated Effluents from Jaipur Dairy." *International Journal of Chemical, Environmental and Pharmaceutical Research 4,* (2): 55-58.

Smol, M., C. Adam, and M. Preisner (2020). "CE Model Framework in the European Water and Wastewater Sector." *Journal of Material Cycles and Waste Management, 22:* 682-697.

Stanchev, P., V. Vasilaki, D. Egas, J. Colon, S. Ponsá and E. Katsou (2020). "Multilevel Environmental Assessment of the Anaerobic Treatment of Dairy Processing Effluents in the Context of Circular Economy."*Journal of Cleaner Production, 261*: 121142.

Voulvoulis, N. (2018). "Water Reuse from a CE Perspective and Potential Risks from an Unregulated Approach." *Current Opinion in Environmental Science & Health, 2:* 32-45.

Wit, Marc de, Jelmer Hoogzaad, Shyaam Ramkumar, Harald Friedl and Annerieke Douma (2018). *The Circularity Gap Report* (*Amsterdam: Circle Economy*). https://policycommons.net/artifacts/3370340/ad6e59_c497492e589c4307987017f04d7af864/4169004/20.500.12592/296r56

Worrell, W.A., P.A. Vesilind, and C. Ludwig (2016). *Solid Waste Engineering: A Global Perspective,* 3rd edn. Boston, MA: Cengage Learning.

WWAP (United Nations World Water Assessment Programme) (2017). *The United Nations World Water Development Report 2017: Wastewater: The Untapped Resource,* Paris: UNESCO.

Zijp, M.C., S.L. Waaijers-vanderoop, R. Heijungs, M.L.M. Broeren, R. Peeters, A. Van Nieuwenhuijzen, L. Shen, E.H.W. Heugens, L. Posthuma (2017). "Method Selection for Sustainability Assessments: The Case of Recovery of Resources from Waste Water". *Journal of Environmental Management, 197*: 221-230. https://doi.org/10.1016/j.jenvman.2017.04.006

13

Green Pesticides in Agricultural Sustenance

An Attempt to Inch Closer to SDG-2

Neha Sharma, Sushma Gautam, Sonali Bhambhani and *Jyoti Sharma*

Introduction

Tomato (*Lycopersicon esculentum* Mill.) is a member of solanaceae (nightshade) family (Table 13.1). Although, it has its roots of origin in Peru, South America but consumed globally. Tomato has been considered as *"green pharmacy"* and is validated as a powerhouse of lycopene, potassium, iron, folate, vitamin C and antioxidants such as beta-carotene and phenolic compounds such as flavonoids, hydroxycinnamic acid, chlorogenic, homovanillic acid, and ferulic acid. Additionally, it is known to alleviate chronic ailments like cancer (lung, prostate, stomach, cervical, breast, oral, colorectal, oesophageal, pancreatic), hypertension, oedema, kidney and liver diseases. Moreover, it plays an active role in detoxification accredited to presence of chlorine and sulphur in fruit. Globally, tomato is the second most cultivated and consumed crop following potato. In 2020, the global production of tomato was 186 million tons (Collins *et al.*, 2022, Shukla *et al.*, 2013).

Table 13.1: Classification of Tomato

Kingdom	**Plantae**
Order	Solanales
Family	Solanaceae
Genus	*Lycopersicum*
Species	*Esculentum*

Growth and development of plant are dependent both biotic and abiotic factors. To accommodate with harsh conditions, plant use various morphological, physiological, biochemical adaptations (Nejat and Mantri, 2017; Bai and Lindhout, 2007). Biotic stress includes pathogenic manifestation caused by bacteria, fungi, virus, nematodes and insects (200 pests are reportedly detrimental to tomato crop). Anthropogenic activities have led to climate change, which in turn contributes to diversification in pathogen habitat culminating into evolution of enhanced virulence (Harvell *et al.*, 2002; Garrett *et al.*, 2006). Tomato is most commonly infested by lepidopterans, coleopterans, dipterans, mites, whitefly, aphids and thrips (Kumar *et al.*, 2022). It has been estimated that arthropods are known to destroy an estimated 18-20 per cent of annual crop with an estimate value of US$470 billion with a prediction of losses being significantly higher in Asia and Africa (Arora, 2017). Figure 13.1 corroborates with commonly encountered pests by *Lycopersicon esculentum* Mill.

The high prevalence of diseases and pests in tomato has resulted in indiscriminate use of pesticides as control agents. Unprecedented use of insecticides has resulted in waning of natural enemies and has significantly contributed to bio-accumulation and bio-magnification of these xenobiotic compounds leading

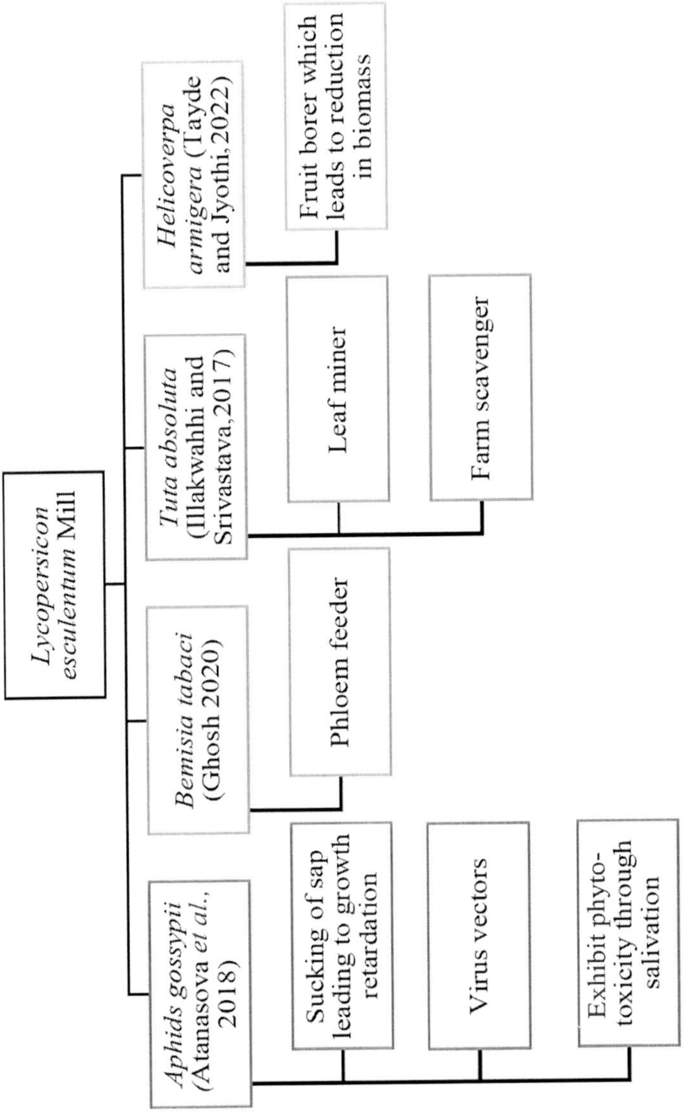

Fig. 13.1: Pest manifestation on *Lycopersicon esculentum* Mill.

to environmental hazards (Khan and Damalas, 2015, Sharma *et al.*, 2019). Imparting host plant resistance has been sought as the most cost-effective method, but such resistance is deemed unstable attributed to insect demographic pressure being exerted on the host, resulting in the evolution of novel biotypes and resistance breakdown (Kumar *et al.*, 2022). Several health issues have been inadvertently reported attributed to unprecedented use of pesticides in farmer centric community (Bakhsh *et al.*, 2017; Kumela *et al.*, 2019). Noteworthy is the fact, around 385 million cases of Unintentional Acute Pesticide Poisoning (UAPP) in farmer centric community with almost 11,000 cases leading to lethal modalities (Boedeker *et al.*, 2020) including reproductive inhibition, hormonal disorder, immune system disorder, and death have been reported (Hassaan and El Nemr, 2020).

Understanding this critical gap analysis and an approach to validate botanicals by *green route* is known to be non-toxic and relatively harmless to indigenous insects and have been explored for their role in Integrated Pest Control Programmes (Pandey *et al.*, 2011). A special impetus has been laid to bio-pesticides capable of providing an alternative to the once derived from animals, plants, bacteria and certain minerals. Bio-pesticides are known to exhibit their bio-efficacious potential through a targeted approach (Bhutto *et al.*, 2017). They are effective even in lower doses and tend to *bio-amalgamate* quickly resulting in lower exposures exhibiting less deleterious effects. Until recently, it has been reported that plant extracts act in different ways, viz. Insect Growth Regulators (IGR), feeding deterrents, repellents (Illakwahhi and Srivastava, 2017). An elucidative account pertaining to the same comprehends with the advantageous role of phytochemicals exhibiting pesticidal efficacy *in vitro*. Bio-active ingredients with known efficacies have been accounted in literature. Table 13.2 corroborates the same.

Table 13.2: Phytochemicals as prospective bio-pesticides

Sl. No.	Plant/Plant component	Bioactive compound	Activity exhibited	Insect	Reference
1.	*R. communis* Leaf	Ricinine	Molluscicidal, Larvicidal, Insecticidal	*Lymnaea acuminate* *Anopheles arabiensis*, *Callosobruchus chinensis*, *Culex Quinquefasciatus* Termites	Bhakta and Das (2015)
2.	*A. indica* Leaves, seeds	Azadirachti, Salannin, Meliantriol, and Nimbin	Insecticidal Repellence Antifeedant	Aphids, whitefly, *Amrasca devastans* *P. xylostella* *T. absoluta*	Kumar and Poehling (2006) Mahmood *et al.* (2014) Liang *et al.* (2003) Alam *et al.* (2019)
3.	*C. limon* Peel, leaves	Limonene and linalool	Insecticide Ovicidal-larvicidal Repellent Aphicidal	*T. absoluta* Whitefly *Macrosiphum roseiformis*	Shiragave (2017) Sweetha (2021) Gupta *et al.* (2017)
4.	*Chrysanthemum cinerariifolium* (Trev.) Vis Flower	Pyrethrin	Insecticidal Repellent Acaricide	*Spodoptera littoralis* *Myzus persicae* *Tetranychus-urticae*	Pavela (2009)
5.	*Allium sativum* Clove	Diallyl sulfphide, diallyl di-sulfide, diallyl trisulfide	Insecticidal Larvicidal Repellent	*T. absoluta*	Alam *et al.* (2019)

Some scalable studies presented below are in harmony with the bio-efficacy of medicinal plants subsequent with their phyto-constituents as prospective pesticidal activity.

A recent study validating polyhouse practices for tomato production results into pocket friendly, eco-effective and sustainable way of agriculture has been reported in literature. Maximized output was attained by fertilizers conjointly used as organic and inorganic moieties (Singh *et al.*, 2021).

In line with similar 3 years study reported in Rajasthan based on frontline demonstration experiment (FLDs) comprising hybrid seed, seed treatment and priming and using modern agricultural practices. The approach was found to enhance the yield by 21.02 per cent as a contrast to conventional agricultural practices (Choudhary *et al.*, 2022).

A previous study related to pest infestation and adversities in tomato, highlighted the potential of chemical pesticide and fungicides in causing eco-toxicity, residual effect and evolution of new biotype. The findings were suggestive of development of resistance in host plant against infecting pathogens and insects (Kumar *et al.*, 2022).

Earlier literature review on food and agricultural security considering whitefly management and IPM depicted application of foliar chemicals being helpful in control of sap sucking pest, but also exhibits potential environment deteriorating and bio-accumulating caliber. Thus, to promote sustainable agriculture, the collaboration of biotechnological interventions with IPM strategies could be impactful (Abubakar *et al.*, 2022).

Previously performed field experiments to trace incidents of sap sucking pests, *H. armigera* and *T. absoluta,* reported first appearance of phloem suckers after second week of transplantation. The peak population of *Aphis gossypii* Glover, *Bemisia tabaci* Gennadius and *Frankliniella schultzei* Trybom was observed during 42nd and 44th standard meteorological week

(SMW), respectively. For fruit borer and leaf miner maximum population of phloem suckers was in 41st meteorological week (SMW) (Kachave *et al.*, 2020).

A 2021 based study was conducted on tomato to assess the toxicity and pesticidal properties of Malathion 57 per cent EC, Potegon 20 per cent SP, K.Z oil, and silver nano particles derived from Neem against aphids. Malathion 57 per cent EC showed the maximum efficacy followed by Potegon 20 per cent SP and K.Z. oil. Findings were indicative of promiscuous use of botanicals as pesticides (Allam *et al.*, 2021).

Aphid affected tomato field was tested for efficacy of seven different treatments. Best efficacy was exhibited by Acetamiprid followed by Neem + Spilanthes, Polygonum plant extracts, Spilanthes, tobacco extracts, Neem pesticides and garlic extracts. Bio-pesticides were observed to be non-toxic and biodegradable, thus recommended for sustainable and eco-friendly agricultural practices (Hassaan and El Nemr, 2020).

A 2017 field-based experiment showed maximum density of *Bemisiatabaci* Genn on tomato during mid-February to mid-March and April-November. Pesticidal property of various chemicals and phyto-chemicals were assessed to control population of whitefly. Synthetic pesticides showed the best output followed by botanicals in combination, i.e. Neem-Spilanthes extract the least efficacy was observed in individual application of botanicals. Thus, this study highlighted the synergistic effect of botanicals (Subba *et al.*, 2017).

A previous study validating application of IPM approach for controlling whitefly on tomato plant was conducted in which crop was supplemented with microbial toxins and botanical extracts, i.e. avermectin (Vertimec 1.9 EC), azadirachtin (Neemactin 0.15 EC), *Spilanthes paniculata* floral parts extract and garlic bulb (*Allium sativum*). The synergistic effect of two treatments containing mixture of azadirachtin and Spilanthes with a formulation constituting (2.5 ml + 80 ml/L) and mixture of azadirachtin and garlic (2.5 ml + 80 ml/L) were evaluated and contrasted with the ability of imidacloprid. The observation shows that imidocloprid exhibited best control over whiteflies followed by avermectin and azadirachtin in combination with spilanthes. Individual application of azadirachtin did not yield efficacious output in controlling whiteflies as compared to combination of botanical extracts (Ghosh, 2020).

A confirmatory study of phytotoxic and insecticidal activity of synthetic as well as bio-insecticides on *Bemisia tabaci* concluded the potential of phytochemicals (castor oil, pongamia oil and sesame oil) to replace chemical pesticides (Kumar *et al.*, 2019).

In line with the concept of green pesticides, it was reported that *Neem* leaf extract was the most effective treatment against leaf miner followed by garlic cloves, lemon, Bishkathali and Mahogany seed extract (Shiragave, 2017).

A 2012-based experimental study on efficacy of fermented plant extract against sapsuckers reported that better performance of botanicals (leaves of *Azadirachta indica* A. Juss and wild garlic cloves) was observed when used in mixed proportions, illustrious of synergism (Nzanza and Mashela, 2012).

Challenges and Future Prospects

Research community has been working intensively on concussion caused by pest infestation in crop from decades not to forget the havoc caused by east African locusts (Devi, 2020). However, in making of greener, cleaner alternatives of chemical pesticides several plants were tested for their capability of pest population regulation in agricultural field. The effectiveness of botanicals may contribute to the expansion and development of organic farming. Several aspects of botanical utilization have not been completely studied yet. Henceforth, below listed points require urgent attention for further growth of organic sector.

(i) Detailed mechanism of pest control by botanical needs to be studied (Cui *et al.*, 2020).

(ii) Currently, the effectiveness of an extract is determined by observing the death rate of target pest. However, some botanicals are found to regulate the reproduction and growth of target pest. Thus, there is need to diversify indicators of efficiency of plant-based bio-pesticides (Cui *et al.*, 2020).

(iii) Although botanicals are known to be harmless to non-target insects. However, researcher tend to ignore to test the effect of bio-pesticide on non-target insects, which compromise the diversity of insects (Papanastasiou *et al.*, 2017). Impact of different pesticides is highlighted in Table 13.3 (Guo *et al.*, 2024).

Table 13.3: Commonly used pesticides (chemically originated and plant derived) with their impact analysis

Category	Commonly used	Advantages	Setbacks
Conventional chemical pesticides	Bifenthrin, benzoylate insecticide, dichloro-diphenyl, trichloro-ethane	• Quick action • Easy to use • Broad-spectrum (can target multiple pest) • Long term impact	• Non-biodegradable • Bio-accumulation and bio-magnification • Lethal to non-target organism including human
Plant-based	Pyrethrum, matrine, nicotine	• Biodegradable • Does not cause ecotoxicity • Cost-effective • Trend to cause lesser resistance among pest; complex active ingredients	• Require huge amount to raw material, active part in plants is in low concentration • Slow working • Effected by regional and seasonal factors • Cannot be stored for long time

Conclusion

Indian agricultural landscape has been relying on chemical and synthetic pesticides in order to abate pest associated losses. The same has been proving detrimental to both abiotic and biotic resources afflicting adverse health issues as reflected in review. With an attempt to overcome eco-toxicological manifestation resonated by pesticides, a green route to obliterate menace caused by xenobiotic pesticides has been documented in literature. A wide array of medicinal plants, their bio-active components have been known to possess pesticidal activity. By throughput screening, such plants are screened to formulate *novel bio-therapeutics* aimed at Integrated Pest Management (IPM) by an inclusive *green* roadmap. This elucidative account is opined to culminate into a better understanding of dynamics of plant secondary metabolites and antagonistic pests. Appropriate integration of numerous ecologically sound treatments like botanicals with minimum use of pesticides may provide a superior management of disease.

Acknowledgements

The authors would like to thank Department of Bio-sciences and Technology, Maharishi Markandeshwar Engineering College, Maharishi Markandeshwar (Deemed to be University), Mullana-Ambala for letting us implement this book chapter with throughput provision of resources.

REFERENCES

Abubakar, M. *et al.* (2022). "Whitefly (*Bemisiatabaci*) Management (WFM) Strategies for Sustainable Agriculture: A Review." *Agriculture 12*(9): 1317.

Alam, M.J. *et al.* (2019). "Bio-Efficacy of Bio-Pesticides against Tomato Leaf Miner, *Tuta absoluta*, a Threatening Pest of Tomato". *Journal of Bioscience and Agriculture Research, 22*(2): 1852-1862. https://doi.org/10.18801/jbar.220219.229

Allam, R.O.H., A.M. Fahmy, and I.R.M. El-Zoghby (2021). "Efficacy of Some Insecticides and Alternatives against *Aphis gossypii* Glover in Tomato Crop and Their Effect on Soil Fertility." *JPPP, 12*(11): 827-832.

Arora, R. (2017). *Breeding Insect Resistant Crops for Sustainable Agriculture.* Singapore: Springer Nature.

Atanasova, D., D. Ganchev, and N. Nenov (2018). "Efficacy of Some Plant Essential Oils against Cotton Aphid, *Aphis gossypii* Glover (Hemiptera: Aphididae) under Laboratory Conditions." *J. Agric. Sci., 1*: 10-16.

Bai, Y. and P. Lindhout (2007). "Domestication and Breeding of Tomatoes: What Have We Gained and What Can We Gain in the Future?" *Annals of Botany, 100*: 1085-1094. https://doi.org/10.1093/aob/mcm150

Bakhsh, K. *et al.* (2017). "Health Hazards and Adoption of Personal Protective Equipment during Cotton Harvesting in Pakistan." *Science of the Total Environment, 598*: 1058-1064. https://doi.org/10.1016/j.scitotenv.2017.04.043

Bhakta, S. and S.K. Das (2015). "In Praise of the Medicinal Plant *Ricinus communis* L.: A Review." *Global Journal of Research on Medicinal Plants & Indigenous Medicine, 4*(5): 95-105.

Bhutto, Z.A. *et al.* (2017). "Integrated Pest Management of Okra Insect Pests." *International Journal of Fauna and Biological Studies, 4*(3): 39-42.

Boedeker, W. *et al.* (2020). "The Global Distribution of Acute Unintentional Pesticide Poisoning: Estimations Based on a Systematic Review". *BMC Public Health, 20*(1): 1-19. https://doi.org/10.1186/s12889-020-09939-0

Choudhary, M.L., R.L. Soni, and B.L. Roat (2022). "Impact of Front-line Demonstration on Yield and Economics of Tomato (*Solanum lycopersicum* Mill.) in Tribal Area of Dungarpur District of Rajasthan." *J. Pharm. Innov., 11*(6): 612-661.

Collins, E.J., C. Bowyer, A. Tsouza and M. Chopra (2022). "Tomatoes: An Extensive Review of the Associated Health Impacts of Tomatoes and Factors That Can Affect Their Cultivation." *Biology, 11*(2): 239. https://doi.org/10.3390/biology11020239

Cui, K. *et al.* (2020). "Comparison of Transcriptome Profiles of the Fungus *Botrytis cinerea* and Insect Pest *Bradysiaodoriphaga* in Response to Benzothiazole." *Frontiers in Microbiology, 11*: 1043. https://doi.org/10.3389/fmicb.2020.01043

Devi, S. (2020). "Locust Swarms in East Africa Could Be 'a Catastrophe'". *The Lancet, 395*(10224): 547.

Garrett, K.A., S.P. Dendy, E.E. Frank, M.N. Rouse, and S.E. Travers (2006). "Climate Change Effects on Plant Disease: Genomes to Ecosystems". *Annual Review of Phytopathology, 44*(1): 489-509. https://doi.org/10.1146/annurev.phyto.44.070505.143420

Ghosh, S.K. (2020). "Environmentally Sound Approach for Management of Tomato Whitefly (*Bemisiatabaci* Genn.)." *J. Entomol. Zool., 8*(6): 814-818.

Guo, C. *et al.* (2024). "Advances in Research and Utilization of Botanical Pesticides for Agricultural Pest Management in Inner Mongolia, China." *Chinese Herbal Medicines, 16*(2): 248-262.

Gupta, G. *et al.* (2017). "Aphicidal Effects of Terpenoids Present in *Citrus limon* on *Macrosiphum roseiformis* and Two Generalist Insect Predators". *Journal of Asia-Pacific Entomology, 20*(4): 1087-1095.

Harvell, C.D., C.E. Mitchell, J.R. Ward, S. Altizer, A.P. Dobson, R.S. Ostfeld and M.D. Samuel (2002). "Climate Warming and Disease Risks for Terrestrial and Marine Biota." *Science, 296*: 2158-2162. https://doi.org/10.1126/science.1063699

Hassaan, M.A. and A. El Nemr (2020). "Pesticides Pollution: Classifications, Human Health Impact, Extraction and Treatment Techniques." *Egypt. Journal of Aquatic Research, 46*(3): 207-220. https://doi.org/10.1016/j.ejar.2020.08.007

Illakwahhi, D.T. and B.B.L. Srivastava (2017). "Control and Management of Tomato Leafminer—*Tuta absoluta* (Meyrick) (Lepidotera, Gelechiidae): A Review." *IOSR J. Appl. Chem. 10*: 14-22. https://doi.org/10.9790/5736-1006011422

Kachave, D.R., M.M. Sonkamble, and S.K. Patil (2020). "Population Dynamics of Major Insect Pests Infesting to Tomato, *Lycopersicon esculentum* (Miller)." *JPP, 9*(3): 344-348.

Khan, M. and C.A. Damalas (2015). "Farmers' Willingness to Pay for Less Health Risks by Pesticide Use: A Case Study from the Cotton Belt of Punjab, Pakistan." *Sci. Total Environ, 530*: 297-303. https://doi.org/10.1016/j.scitotenv.2015.05.110

Kumar, Dharamraj, Anjali Rani, Jaswant Prajapati, Sudipta Mahato, Narendra Pratap Verma, Astha Vishwaraj, Suman Kumar and Devesh Sanjay Pardhi (2022). "Breeding for Biotic Stresses Resistance in Tomato." *A.J. Pharma Innov, 11*(5): 316-321.

Kumar, P. and H.M. Poehling (2006). "Persistence of Soil and Foliar Azadirachtin Treatments to Control Sweetpotato Whitefly *Bemisiatabaci* Gennadius (Homoptera: Aleyrodidae) on Tomatoes under Controlled (Laboratory) and Field (Netted Greenhouse) Conditions in the Humid Tropics." *Journal of Pest Science, 79*: 189-199.

Kumar R. *et al.* (2019). "Insecticidal Activity of Botanical Oils and Other *Neem*-based Derivatives against Whitefly, *Bemisiatabaci* (Gennadius) (Homoptera: Aleyrodidae) on Cotton". *International Journal of Tropical Insect Science, 39*: 203-210. https://doi.org/10.1007/s42690-019-00027-4

Kumela, T. *et al.* (2019). "Farmers' Knowledge, Perceptions, and Management Practices of the New Invasive Pest, Fall Armyworm (*Spodoptera frugiperda*) in Ethiopia and Kenya." *International Journal of Pest Management, 65*(1): 1-9. https://doi.org/10.1080/09670874.2017.1423129

Liang, G.M., W. Chen, and T.X. Liu (2003). "Effects of Three *Neem*-based Insecticides on Diamondback Moth (Lepidoptera: Plutellidae)". *Journal of Crop Protection, 22*(2): 333-340. https://doi.org/10.1016/S0261-2194(02)00175-8

Mahmood, K. *et al.* (2014). "Effects of Biopesticides against Jassid [*Amrascadevastans* (dist.)] and White Fly [*Bemisiatabaci* (genn.)] on Okra". *International Journal of Biology and Biotechnology, 11*(1): 161-165.

Nejat, N. and N. Mantri (2017). "Plant Immune System: Crosstalk between Responses to Biotic and Abiotic Stresses the Missing Link in Understanding Plant Defence." *CIMB,* 23(1): 1-16. http://dx.doi.org/10.21775/cimb.023.001

Nzanza, B. and P.W. Mashela (2012). "Control of Whiteflies and Aphids in Tomato (*Solanum lycopersicum* L.) by Fermented Plant Extracts of *Neem* Leaf and Wild Garlic." *African Journal of Biotechnology, 11*, 94: 16077-16082. https://doi.org/10.5897/AJB12.775

Pandey, A. *et al.* (2011). "Formulation and Evaluation of Antibacterial and Antifungal Activity of Herbal Gel Containing *Aloe vera, Azadirachta indica* and *Lycopersicon esculentum* Seed Extract." *Research Journal of Pharmacy and Technology, 4*(4): 553.

Papanastasiou, S.A. *et al.* (2017). "Toxic and Hormetic-like Effects of Three Components of Citrus Essential Oils on Adult Mediterranean

Fruit Flies (*Ceratitis capitata*)." *PLOS One,* 12(5): e0177837. https://doi.org/10.1371/journal.pone.0177837

Pavela, R. (2009). "Effectiveness of Some Botanical Insecticides against *Spodoptera littoralis* Boisduvala (Lepidoptera: Noctudiae), *Myzus persicae* Sulzer (Hemiptera: Aphididae) and *Tetranychusurticae* Koch (Acari: Tetranychidae)". *Plant Protection Science, 45*(4): 63.

Sharma, Anket, Vinod Kumar, Babar Shahzad, Mohsin Tanveer, Gagan Preet Singh Sidhu, Neha Handa, Sukhmeen Kaur Kohli, Poonam Yadav, Aditi Shreeya Bali, Ripu Daman Parihar, Owias Iqbal Dar, Kirpal Singh, Shivam Jasrotia, Palak Bakshi, M. Ramakrishnan, Sandeep Kumar, Renu Bhardwaj & Ashwani Kumar Thukral. (2019). "Worldwide Pesticide Usage and Its Impacts on Ecosystem." *SN Applied Sciences, 1*: 1446. https://doi.org/10.1007/s42452-019-1485-1

Shiragave, P.D. (2017). "Biochemical Screening and Larvicidal and Ovicidal Activity *Citrus limon* (L.) Burm. F. Leaves against *Helicoverpa armigera* (Hub.)." *Journal of Entomology and Zoology Studies, 5*(1): 885-888.

Shukla, Priya, Katyani Bajpai, Shalini Tripathi, Sunil Kumar and Girendra Kumar Gautam (2013). "A Review on the Taxonomy, Ethnobotany, Chemistry and Pharmacology of *Solanum Lycopersicum* Linn." *IJCPS 1*(8): 521-527.

Singh, M. *et al.* (2021). "An Economic Analysis of Tomato Cultivation under Polyhouse Condition." *Current Journal of Applied Science and Technology, 40*(44): 39-43.

Subba, B. *et al.* (2017). "Population Dynamics of White Fly (*Bemisiatabaci* Genn.) Infesting Tomato (*Lycopersicon esculentum* L.) and Their Sustainable Management Using Bio-Pesticides." *J. Entomol. Zool. 5*(3): 879-883.

Sweetha, G. (2021). "Is Lemon Peel Responsible for Controlling White Fly? A Review Article." *International Journal of Scientific Development and Research, 6*(5): 397-399.

Tayde, A.R. and A. Jyothi (2022). "Field Efficacy of Some Insecticides and *Neem* Products against Tomato Fruit Borer [*Helicoverpa armigera* (Hubner)] in Tomato." *Int. J. Plant Soil Sci., 34*(23): 299-303. https://doi.org/10.9734/ijpss/2022/v34i2331593

14

Sustainable Agriculture Using Modern Technologies

Alfreddie Rani and *R.K. Josmee*

Introduction

World agriculture is now facing several interconnected challenges, such as accelerated population growth, climate change, dwindling natural resources, and increasing food production demand. These strains require efficient and sustainable farming techniques that can raise productivity with less environmental deterioration. Here, precision farming has become a central methodology, utilising cutting edge technology like Global Positioning Systems (GPS), Geographic Information Systems (GIS), and remote sensing in order to facilitate site specific and data oriented crop management practices.

One of the most remarkable breakthroughs in this regard is the use of Unmanned Aerial Vehicles (UAVs), popularly known as drones. Originally created for military use, UAVs have evolved to find civilian use in numerous applications, with agriculture being one of the most successful. Their ability to capture high-resolution, real-time images under a variety of environmental conditions makes them extremely useful for crop monitoring, yield forecasting, and input optimization. In particular, UAVs are used to measure plant vigour, identify biotic and abiotic stress factors, and control irrigation, fertilization, and pesticide application more accurately.

Armed with advanced sensors, e.g. multispectral, hyperspectral, thermal, and RGB camera, drones allow for in-depth examination of crop health status. These technologies allow for the early identification of pest infestations, diseases, water stress, and nutrient deficiencies, thus enabling timely and effective agronomic management. Empirical research has established that UAVs often surpass satellite and ground based systems for site specific crop health monitoring owing to their high spatial and temporal resolution.

New advances in technology have further increased the application of UAVs with the addition of artificial intelligence (AI) and Internet of Things (IoT). With this, data processing have been automated, crop classification made smart, anomaly detection achieved, and precision spraying activities carried out. This way, UAVs make a significant input towards enhancing agricultural productivity, resource efficiency, and sustainability.

Nonetheless, the extensive application of UAV technology in agriculture is still hampered by several factors. Main limitations are the restricted battery life, low payload capacity, high costs of implementation and maintenance, as well as regulatory impediments concerning airspace management and flight operations. These limitations are especially severe in remote and resource poor areas, accentuating the necessity for policy interventions and technological modifications to promote increased accessibility.

This chapter explores the use of UAV remote sensing technology in monitoring crop health in Mawkriah, a rural agricultural setting that is known for geographical and climatic heterogeneity. The case study chosen provides a rich opportunity to evaluate the field applicability, advantages, and drawbacks of precision agriculture supported by drones in real field settings. Through the use of UAVs that are fitted with multispectral sensors, the study seeks to assess vegetation indices, detect crop stress areas, and add to the larger conversation regarding how UAVs can be used to improve decision making and sustainability in agricultural systems.

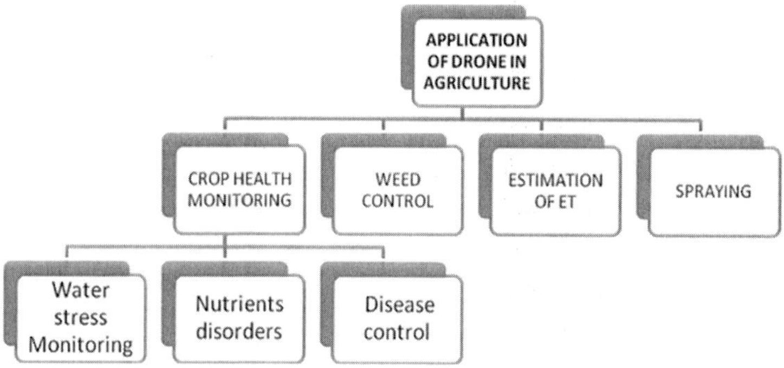

Fig. 14.1: Different applications of drone in agriculture.

Study Area

This study was conducted in Mawkriah, a rural agricultural region characterized by its heterogeneous topography and varied climatic conditions. Mawkriah East is a village situated in the Mylliem Subdivision of the East Khasi Hills District in Meghalaya, India. It lies approximately 9 kilometres from the subdivisional headquarters in Mylliem and about 17 kilometres from the district headquarters in Shillong. Located at Longitude 91.786376°N and Latitude 25.503768°E with an estimated area of 135 hectares (study area).

The site was selected due to its representative mix of smallholder plots, diverse cropping patterns, and practical relevance for assessing the real world applicability of drone based crop monitoring. The predominant crop selected is cauliflower, known for its exceptional freshness, superior quality, and larger size, grown with minimal chemical fertilisers, offering great value at competitive prices and cultivated under both rainfed and irrigated conditions.

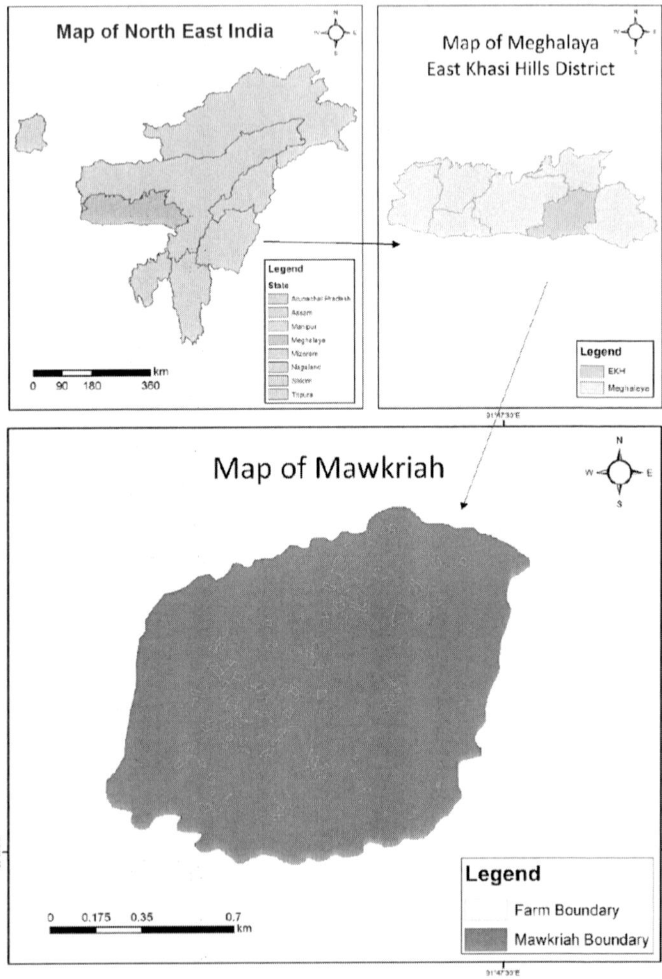

Fig. 14.2: Map of Study Area.

Methodology

Household Survey and Farmer Registration

To contextualise the remote sensing data and ensure effective field validation, a household level economic survey was conducted prior to UAV deployment. This survey captured demographic and farming related information from individual households. Data collected included:

- Farmer's full name.
- Contact number.
- Farm size.
- Type of crops grown.
- Seasonal cropping patterns.
- Input use and irrigation practices.

A semi-structured questionnaire was used to collect data, and we conducted field interviews and documented the responses in a database. This set of data provided an initial socio-economic context that amended the household study, while also being used for follow-up questions raised during the field validation and interpretation of the NDVI data.

Plot Demarcation and Georeferencing: After the survey was conducted, we physically demarcated each agricultural plot using GPS. Each of the farmers' fields was mapped, and each field was assigned a unique identifier with an accompanying record in the household survey. Their geospatial boundaries were also digitised in GIS software (ArcGIS) to align properly with UAV imagery. Vegetation indices were analysed by farm plot and further correlated with indicated socio-economic conditions.

Data Acquisition, UAV: To identify plant health trends/ archives for the agricultural plots, multispectral imagery from a UAV (using a multispectral camera to collect imagery from the visible and NIR bands) was collected. This drone platform (M1000 Assembled) was flown at an altitude of 60,100 metres; the researcher needed to balance between spatial resolution and coverage. The UAV flights were flown when visual conditions were established (clear sky), to control atmospheric interference and assure consistency in the captured images. The survey area covered 135 hectares, and the work was conducted at a relevant growth stage of the crops studied, specifically, when the crops were in the vegetative and reproductive periods (March and October), when plant health changes are detectable.

The multispectral sensor captured at least four spectral bands (Blue, Green, Red, and Near Infrared), which are essential for calculating vegetation indices such as the Normalized Difference Vegetation Index (NDVI).

Image Processing and NDVI Generation

Postflight, the collected imagery was processed using Pix4D mapper and Agi soft Meta shape, where images were georeferenced, Ortho mosaic, and radiometrically corrected. NDVI was then calculated using the standard formula:

$$NDVI = \frac{NIR - Red}{NIR + Red}$$

Fig. 14.3: NDVI formula.

This index is widely used as a proxy for vegetation health, with values ranging from 1.0 to +1.0. In agricultural applications, NDVI values closer to +1 typically indicate dense, healthy vegetation, while lower values suggest sparse, stressed, or senescent crops.

Crop Health Classification

Once NDVI values were computed, the resulting index maps were used to classify crop health across the study area. A threshold based classification was applied, where NDVI values were segmented into the following categories:

- Dead plants or Inanimate objects (10)
- Unhealthy (00.33)
- Moderately healthy (0.33-0.66)
- Healthy plant (0.661)

In general, higher values indicate more vegetation or healthy crops, while lower values indicate less healthy plants or crops that are not growing well. An NDVI value between 0 and 1 is commonly considered optimal for crop growth; values less than 0 indicate areas with too little vegetation for photosynthesis to occur effectively (such as during drought). Negative NDVI corresponds to water.

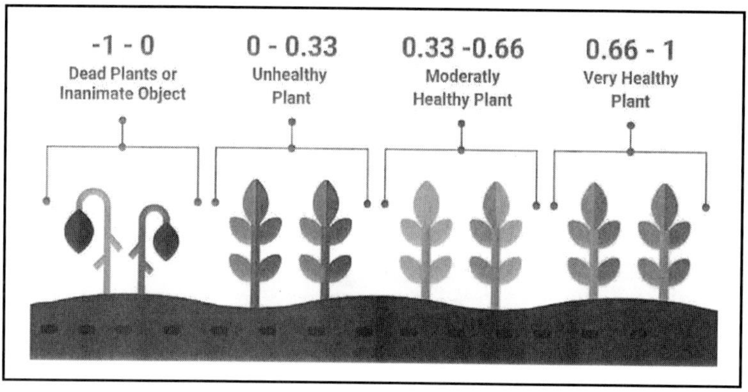

Source: https://cropin.com/blogs/ndvinormalizeddifferencevegetationindex.html

Fig. 14.4: Normalized Difference Vegetation Index.

These thresholds were established according to field observations and literature reviews, modified according to local crop conditions. Ground truthing was conducted simultaneously with UAV surveys and through ground surveys to verify remote sensing data, relying on visual inspection and leaf chlorophyll readings taken at selected sample plots.

Analysis and Interpretation

The NDVI based health maps were analysed to provide an identification of stress and vigour spatial patterns MS. The stress and vigour spatial patterns were then interpreted according to the nature of field conditions observed in the field: pest damage, water stress and nutrient deficiencies allowed for a better understanding of the correspondence of remotely sensed data to actual crop health dynamics in Mawkriah. The study also examined the ways in which NDVI trends varied between different plots and growth stages, more information on the effectiveness of farm management practices and where such practices may need targeted focus.

Fig. 14.5: Flowchart of methodology.

Research Problem

Declining Trends in Cauliflower Farmers in Mawkriah, East Khasi Hills

A field study was conducted in Mawkriah whereby a total of 64 farmers were interviewed about the declining rate of cauliflower plantations in the area. Mawkriah used to be the main supplier of Cauliflower, which also boasts one of the finest in the region, accounting for roughly 42 per cent in the market share. However, with urbanization and a lack of proper incentives from the government, there has been a drastic decline in the number of cauliflower farmers per household. Data collected showed that roughly 52 farmers have turned to other crops or have sold their land to look for other opportunities to improve their livelihood.

This decline poses a threat to the local market, resulting in dependency on imports, which raises the market prices and affects the people of the region.

With this new drone technology coupled with scientific inputs, farmers can restore their faith in advocating cauliflower farming.

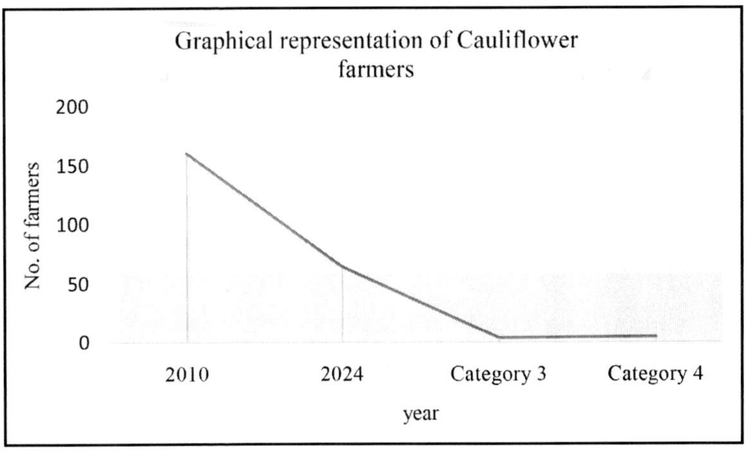

Fig. 14.6: Graphical figure of cauliflower farmers.

Data Integration

1. *Geospatially Referencing Survey Plots*

GPS tagged field locations, which were collected during field work, were associated with survey responses to link the survey responses to specific farm plots. These field locations were then overlaid on the Orth mosaic created by the UAV images and the survey responses were matched with spatial data.

2. *NDVI Analysis and Health Classifications*

The NDVI values were classified into thresholds to identify the following: Healthy vegetation (high NDVI), Zones of moderate stress, and severely stressed areas. The classifications were cross-validated using farmers' observations noted in the surveys.

3. *Cross-validation and Thematic Association*

Select UAV generated metrics (i.e. NDVI) were statistically analysed in comparison to: Recorded pest or disease incidents, fertiliser application, and observed variations in yield. Strong associations were identified, for instance, in intervals where farmers recorded water management issues (poor drainage, and/or lack of irrigation), which spatially corresponded with low NDVI areas. Likewise, the findings regarding disease prone fields follow the signature of stressed vegetation observed in the drone mappings.

Generating Results

The findings were ascertained by synthesizing the spatial and thematic data gathered from UAV based remote sensing in addition to the ground truthing data produced from 64 household surveys. This combination allowed for not only visual identification of potential crop health issues but also determination of the causes and localized manifestation of agricultural stress reported by farmers themselves.

(*a*) *NDVI-based Health Zoning High Resolution*: NDVI-based Health Zoning High Resolution multispectral images collected through drone UAVs were processed to derive NDVI for all the plots surveyed. NDVI values were separated into three zones:

- Low NDVI: Unhealthy (00.33).
- Moderate NDVI: Moderately healthy (0.33-0.66).
- High NDVI: Healthy plant (0.661).

The NDVI maps were overlaid onto farm plots with GPS tags taken from the household survey, allowing the association of vegetation health patterns with the individual farmers' fields.

(*b*) *Cross-referencing Farmer Observations*: Each NDVI classified zone was cross-referenced with survey responses documenting farmers' field based observations, which included:

- Location of wilting.
- History of pest invasions or fungal outbreaks.
- Fertilizer application patterns and irrigation timetable etc.

For example:

- Low NDVI zones aligned with where farmers reported postponed fertiliser application or poor irrigation.
- Moderate NDVI zones aligned with integrated plots where farmers reported only "some" pest damage and mitigating action had been initiated.

- Other instances emerged where farmers believed their crops were healthy, but the UAV data revealed incipient stress, providing feedback loops for timely responses.

(*c*) *Time Series Analysis and Interventions*: Our multiple UAV flights over two crop cycles enabled us to temporally compare NDVI values for all the assessing plots. From these data, we could assess:

- Positive improvement trends in the fields where changes had induced action (for example: pest control, or timed irrigation).
- Declining rates in plots with chronic issues (for example, poor drainage or degraded soil).

Farmers with data informed interventions led to NDVI improvements in the range of 15-20 per cent, which were consistent with farmer reported yield increases. This indicated that before and after impact of the monitoring approach using UAV guided data informed interventions with the inclusion of farmer said inputs.

(*d*) *Creation of Composite GIS Layers:* All UAV and soil survey data were georeferenced and layered using GIS to create thematic maps depicting the:

- NDVI class by plot.
- Reported management practices (fertilizer type, irrigation method).
- Reported constraints (waterlogging, pest infestation zones).
- Yield reported by farmers.

These thematic maps provided an opportunity to observe patterns at the village scale clusters of vulnerability (e.g. allotments that performed poorly and were located near drainage lines) and areas of resilience were observed.

Results

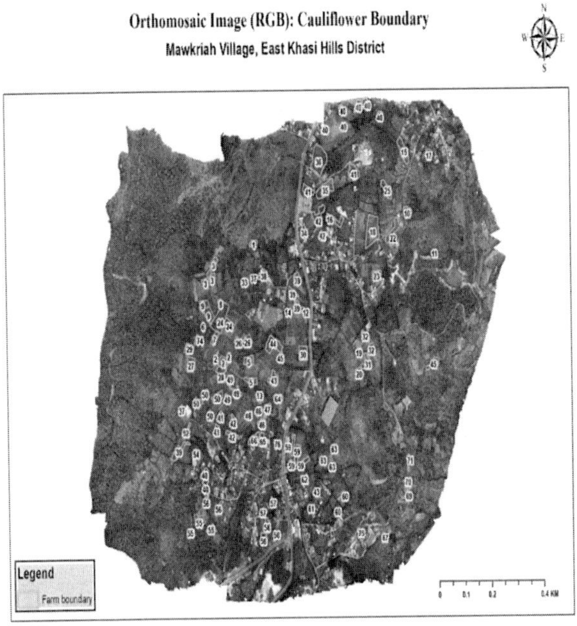

Fig. 14.7: Ortho mosaic Image of Mawkriah (Captured by M1000) along with farm Boundaries.

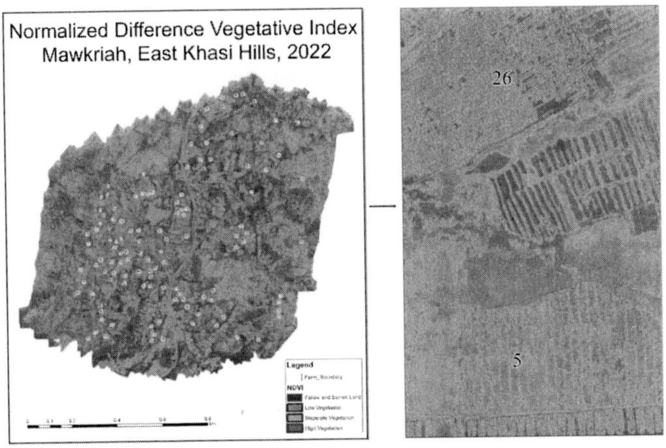

Fig. 14.8: An NDVI map showing vegetation stress across the study area.

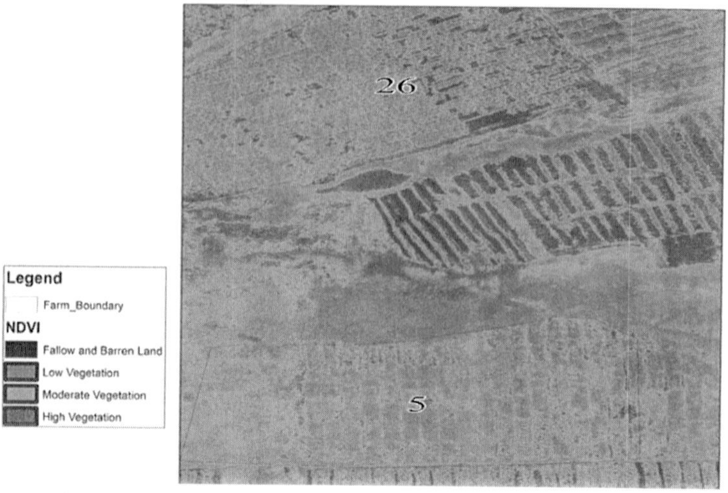

Fig. 14.9: A particular plot showing both low vegetative stress and moderate stress.

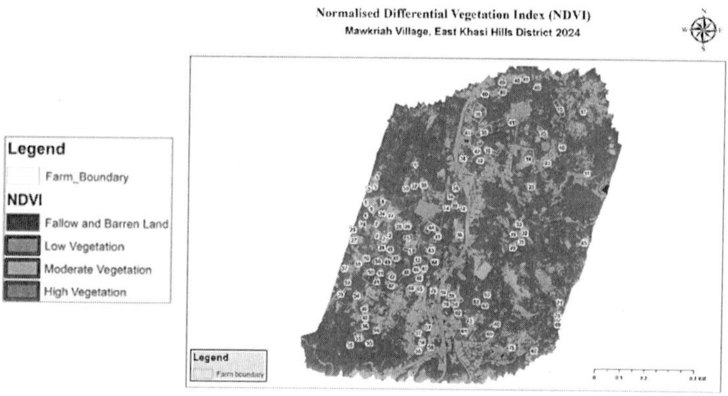

Fig. 14.10: Improvement of vegetation stress in the study area after advisory inputs to farmers.

Discussion

Summary of Data Integration and Baseline Results

This study in Mawkriah East utilized drone based multispectral data alongside socio-economic and field level information collected through structured household surveys with 64 farmers. The baseline NDVI maps presented evidence of a range of vegetation health (NDVI) patterns in the plots, where about 71 per cent of the plots were identified under moderate to high stress levels. These findings closely represented the farmer reported issues of unfavourable soil conditions, irregular irrigation, or the presence of pests.

The initial steps of cross validation between UAV data and farmers' survey responses confirmed that remote sensing indicators were indeed similar to farmers' experiential knowledge. For example, the low NDVI identified on the fields corresponded to the area where farmers reported low levels of crop vigour based on their observation of nutrient imbalance or water shortage.

Adviser Intervention and Engagement of Farmers

After the preliminary data analyses, we held advisory sessions and data sharing workshops with the farmers who participated in the study. Overall, the drone imagery and NDVI maps were provided to the farmers in a visual and interpretable format.

Two Cropping Season Post Advisory Changes and Measurable Improvements

During the two cropping seasons after the advisory, follow-up UAV surveys and field interviews were conducted to assess changes. The NDVI values taken before and after the crop advisory inputs showed definitive improvement in crop health and coherently logical support for post advisory changes:

- NDVI values for previously stressed plots increased by an average of 0.12, an indication of improved vegetation vigour.

- Yield improvement was between 16 per cent and 32 per cent and was in particular noted among farmers that acted upon the advice given for improving fertilizer use and irrigation efficiency.
- Fertilizer use efficiency was better, as it was measured in fertilisers used at the farm level improved benefits of using fertilizer were based on previous spatial variability maps the farmers were applying less based on need. Farmers noted that, by implementing the spatial mapping of crops, they were reducing their fertilizer use by 10 to 15 per cent without losing productivity.
- Soil fertility indicators, such as organic matter and texture (using basic tests that were conducted with local extension support), improved in treated fields because there was better nutrient management and crop residual practices after the advisory from the initiation of the advisory.

Farmer Attitudes and Behaviour Change

The qualitative data we collected through the final round of surveys indicated positive changes in farmer attitudes regarding technology assisted decision making:

- Farmers reported a better understanding of NDVI maps after they engaged in the second cycle.
- Multiple farmers began using UAV based vegetable health classification as a reason to begin implementing plot-wise management (e.g. targeted watering, localized composting).
- Crucially, 52 per cent of respondents indicated they planned to pay for future drone monitoring due to the value of identifying invisible stress, and want to make less deduction based on health classifications and stress levels.

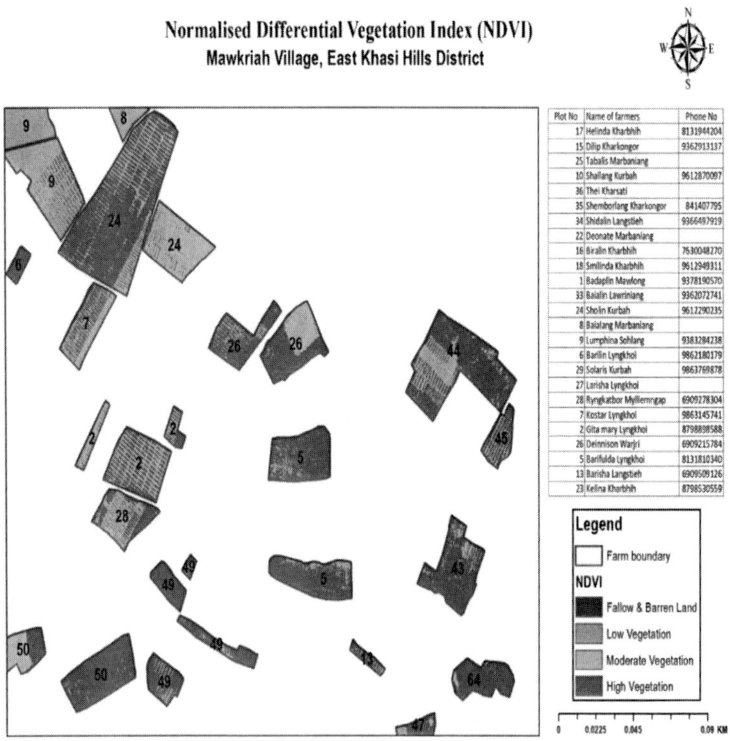

Fig. 14.11: A particular plot showing improvement with the aid of NDVI indices.

Conclusion

An effective means for documenting and monitoring crop health in Mawkriah East, Meghalaya, is demonstrated by integrating planned UAV based remote sensing with household surveys. Combining the two data sets gave value added insights that identified stress zones and generated advisories to monitor crop health locally. Our observations of pre and postintervention assessments demonstrated positive change in NDVI values, crop yield (16-32%), and improved input efficiency (especially fertilizer), which offers operational guidance for the farmers.

In conclusion, our evidence indicates that UAVs' locus based assessment features offered increased precision farming opportunities in smallholder farm systems with the support of participatory data and advisory services. This model could be iterated with UAV technology and provides a scalable model to integrate real time, farm level data to guide sustainable farm management practices, but more importantly, local farm practices can now be bridged closer to UAV technology.

REFERENCES

Ahirwar, S., Swarnkar, R., Bhukya, S. and Namwade, G. (2019). "Application of Drone in Agriculture". *Inter J Curr Microbiol Appl Sci.*, *8*(01).

Barbedo, J.G.A. (2019). "A Review on the Use of Unmanned Aerial Vehicles and Imaging Sensors for Monitoring and Assessing Plant Stress". *Drones*, *3*(2): 40.

Berni, J.A.J., Zarco Tejada, P.J., Suarez, L., Gonzalez Dugo, V. and Fereres, E. (2009a). "Remote Sensing of Vegetation from UAV Platforms Using Lightweight Multispectral and Thermal Imaging Sensors". *International Society for Photogrammetry and Remote Sensing (ISPRS)*.

Beeri, O. and Peled, A. (2009). "Geographical Model for Precise Agriculture Monitoring with Real Time Remote Sensing". *ISPRS Journal of Photogrammetry and Remote Sensing, 64*.

Boori, M.S., Choudhary, K. and Kupriyanov, A.V. (2020). "Crop Growth Monitoring Through Sentinel and Landsat Data-based NDVI Timeseries". *Computer Optics*, *44*(3).

Candiago, S., Remondino, F., DeGiglio, M., Dubbini, M. and Gattelli, M. (2015). "Evaluating Multispectral Images and Vegetation Indices for Precision Farming Applications from UAV Images". *Remote Sensing*, *7*(4).

Faustin, Niyonzima K. (2024). "Applications of Remote Sensing in Precision Agriculture". *Research Output Journal of Engineering and Scientific Research*, *3*(1).

Hafeez *et.al.* (2023). "Implementation of Drone Technology for Farm Monitoring and Pesticide Spraying: A Review". *Information Processing in Agriculture, 10.*

Hegarty Craver, M., Polly, J., O'Neil, M., Ujeneza, N., Rineer, J. and Beach, R.H. *et al.* (2020). "Remote Crop Mapping at Scale: Using Satellite Imagery and UAV Acquired Data as Ground Truth". *Remote Sensing, 12.*

Hoffmann, W.C., Lan, Y., Huang, Y., Wu, W. and Fritz, B.K. (2008). "Development of a Spray System for an Unmanned Aerial Vehicle Platform". *Applied Engineering in Agriculture, 25*(6).

Jeevan *et.al.* (2024). "Crop Health Monitoring Using Drones: A Modern Approach to Precision Agriculture". *Krishi Science, 5*(8).

Khanal, S., Kc, K., Fulton, J.P., Shearer, S. *et al.* (2020). "Remote Sensing in Agriculture Accomplishments, Limitations, and Opportunities". *Remote Sensing, 12*(22).

Lelong, C.C.D., Pinet, P.C. and Poilve, H. (1998). "Hyperspectral Imaging and Stress Mapping in Agriculture: A Case Study on Wheat in Beauce (France)". *Remote Sensing of Environment, 66.*

Mani, P.K., Mandal, A., Biswas, S., Sarkar, B., Mitran, T., Meena and R.S. (2021). "Remote Sensing and Geographic Information System: A Tool for Precision Farming." *Geospatial Technologies for Crops and Soils,* Springer Nature Singapore.

Martos, V., Ahmad, A., Cartujo, P., Ordoñez, and J. Ensuring (2021). "Agricultural Sustainability through Remote Sensing in the Era of Agriculture 5.0". *Applied Sciences, 11*(13).

Sishodia, R.P., Ray, R.L. and Singh, S.K. (2020). "Applications of Remote Sensing in Precision Agriculture: A Review". *Remote Sensing, 12*(19).

Zhou,Y., Lao, C., Yang, Y., Zhang, Z., Chen, H., Chen, Y., Chen, J., Ning, J. and Yang, N. (2021). "Diagnosis of Winter Wheat Water Stress based on UAV Borne Multispectral Image Texture and Vegetation Indices". *Agric. Water Manag., 256.*

Index